Industry 4.0 in SMEs Across the Globe

Industry 4.0 in SMEs Across the Globe

Drivers, Barriers, and Opportunities

Edited by

Julian M. Müller
Nikolai Kazantsev

CRC Press is an imprint of the
Taylor & Francis Group, an **informa** business

First edition published 2022
by CRC Press
6000 Broken Sound Parkway NW, Suite 300, Boca Raton, FL 33487–2742

and by CRC Press
2 Park Square, Milton Park, Abingdon, Oxon, OX14 4RN

CRC Press is an imprint of Taylor & Francis Group, LLC

Library of Congress Cataloging-in-Publication Data

Names: Müller, Julian M., editor. | Kazantsev, Nikolai, editor.
Title: Industry 4.0 in SMEs across the globe : drivers, barriers, and opportunities / edited by Julian M. Müller and Nikolai Kazantsev.
Description: First edition. | Boca Raton, FL : CRC Press, 2022. |
Includes bibliographical references and index.
Identifiers: LCCN 2021036973 (print) | LCCN 2021036974 (ebook) |
ISBN 9780367761905 (hbk) |
ISBN 9780367761912 (pbk) |
ISBN 9781003165880 (ebk)
Subjects: LCSH: Industry 4.0. | Small business.
Classification: LCC T59.6 .I38 2022 (print) | LCC T59.6 (ebook) |
DDC 658.4/038--dc23/eng/20211007
LC record available at https://lccn.loc.gov/2021036973
LC ebook record available at https://lccn.loc.gov/2021036974

ISBN: 978-0-367-76190-5 (hbk)
ISBN: 978-0-367-76191-2 (pbk)
ISBN: 978-1-003-16588-0 (ebk)

DOI: 10.1201/9781003165880

Typeset in Times
by Apex CoVantage, LLC

Contents

Preface

The concept of Industry 4.0 has been increasingly analyzed in the literature; however, to a large extent, the perspective of small and medium-sized enterprises (SMEs) has been restricted to single countries. This book aims to consolidate the experiences of Industry 4.0 implementation in SMEs across different parts of the globe. It describes experiences of Industry 4.0 implementation by providing an in-depth overview of Industry 4.0 in SMEs and covering various national, historical, and geographical settings in nine European countries: Finland, France, Hungary, Italy, Poland, Russia, Lithuania, Serbia, and the UK, complemented by five further countries from around the world: Brazil, China, India, Iran, and the US.

Each chapter describes the national digitalization program, along with barriers, drivers, and opportunities to implement Industry 4.0 in local SMEs. The book subsumes the findings across the 14 countries to identify common themes and clusters of drivers, opportunities, and barriers. It concludes that there is a common path for Industry 4.0 to be better adopted by SMEs to increase industrial competitiveness globally. This path stipulates that Industry 4.0 cannot be implemented in a single firm setting but, rather, in inter-firm settings where the success of conceptually envisioned horizontal, vertical, and end-to-end integration depends on several drivers, such as demand of customers and an SME's role in the supply chain, employee skills and culture of manufacturers, and local digital policy and technology aspects.

Preface

Editors

Julian M. Müller is Professor at Kufstein University of Applied Sciences, Austria, and is Visiting Professor at Jagellonian University Krakow (Poland) and Corvinius University Budapest (Hungary). He completed his PhD at Friedrich-Alexander-University Erlangen-Nürnberg (Germany). His research interests include Industry 4.0, technology and innovation management, supply chain management, sustainability, small and medium-sized enterprises, and business model innovation.

Nikolai Kazantsev is a research associate at the University of Exeter (UK) and a senior lecturer at HSE University (Russia). His PhD from the University of Manchester is devoted to barriers to new supply relationships between SMEs in the context of Industry 4.0, collaboration design methods, and collaborative business models for SMEs. Nikolay's research interests are related to the impacts of digital technologies and new collaborative business models on industries, companies, and individuals.

1 Industry 4.0 Barriers, Drivers, and Opportunities

An Introduction to Lessons Learned from SMEs Worldwide

Julian M. Müller and Nikolai Kazantsev

CONTENTS

1.1 INTRODUCTION

At the Hanover Fair in 2011, the German government announced the concept of "*Industrie 4.0*" as part of its high-tech strategy. The concept, now mostly known as "Industry 4.0," originally aimed to secure the future competitiveness of the German manufacturing industry. It relates to a fourth Industrial Revolution, based on cyber-physical systems (CPS) and the Internet of Things (IoT). CPS aim to resemble and extend the physical world in a virtual one using and generating data. This data generated is shared using the IoT, interconnecting humans, production facilities, and products across the entire value chain. Thus, Industry 4.0 enables horizontal and vertical integration, i.e., across entire industrial value chains, across the entire lifecycle of products, and across several functional departments (Dalenogare et al., 2018; Kagermann et al., 2013; Lasi et al., 2014).

While many technological solutions have been found, Small and Medium-Sized Enterprises (SMEs) still lag behind in several regards, relating to their smaller size, resource base, bargaining power, missing economies of scale, and often acting as suppliers without end customer contact (Horváth & Szabó, 2019; Moeuf et al., 2020; Müller et al., 2018). Hence, SMEs cannot grasp the potentials of Industry 4.0 as large enterprises do and face some distinct barriers against Industry 4.0 (Masood & Sonntag, 2020; Sahi et al., 2020; Stentoft et al., 2020). However, for Industry 4.0 to

DOI: 10.1201/9781003165880-1

unfold successfully, SMEs also need to be integrated within supply chains. Hence, also for large enterprises and entire supply chains, the implementation of Industry 4.0 in SMEs is vital (Birkel & Müller, 2021; Müller et al., 2020, Veile et al., 2020). Therefore, this book, in addition to strengthening understanding of SMEs' implementation of Industry 4.0 around the world, aims to regard SMEs as part of entire supply chains that span the globe, that is, as part of industrial ecosystems. In this regard, SMEs despite their small size, need to be integrated into supply chains and ecosystems to achieve horizontal and vertical integration as a central pillar of Industry 4.0 (Benitez et al., 2021; Hahn, 2020; Schmidt et al., 2020).

1.2 CONTENT OF THIS BOOK

This book includes experiences of Industry 4.0 implementation among SMEs located in 14 countries around the world. Contributions from nine European countries, Finland, France, Hungary, Italy, Poland, Russia, Lithuania, Serbia, and the UK, provide insights from their lessons learned from Industry 4.0 implementation. Those insights are complemented by five further countries from around the world: Brazil, China, India, Iran, and the US. Each chapter briefly describes the context and the digitalization policy adopted in the country first, and highlights the barriers, drivers, and opportunities that companies identified in relation to Industry 4.0.

Following these 14 perspectives on Industry 4.0 implementation, a further chapter explains why SMEs must be integrated better into supply chains. This chapter provides a meta-analysis of several studies with SMEs, as well as their customers and larger supply chain counterparts. Hence, this chapter attempts to extend the SME perspective on Industry 4.0 implementation.

Finally, the last chapter concludes the book, summarizing commonalities and differences among factors influencing Industry 4.0 in SMEs nationally. This chapter further aggregates Industry 4.0 opportunities, illustrating further collaborative possibilities for SMEs.

We want to thank all authors for their valuable contributions to this book, as well as the publisher, who made this publication possible. The interesting insights on the topic of Industry 4.0 and SMEs shall be of benefit for all readers from academia and practice.

Julian M. Müller and Nikolai Kazantsev

REFERENCES

Benitez, G. B., Ferreira-Lima, M., Ayala, N. F., & Frank, A. G. (2021). Industry 4.0 technology provision: The moderating role of supply chain partners to support technology providers. *Supply Chain Management: An International Journal*, *ahead-of-print*(ahead-of-print). https://doi.org/10.1108/SCM-07-2020-0304

Birkel, H. S., & Müller, J. M. (2021). Potentials of industry 4.0 for supply chain management within the triple bottom line of sustainability – a systematic literature review. *Journal of Cleaner Production*, 125612.

Dalenogare, L. S., Benitez, G. B., Ayala, N. F., & Frank, A. G. (2018). The expected contribution of industry 4.0 technologies for industrial performance. *International Journal of Production Economics*, *204*, 383–394.

Hahn, G. J. (2020). Industry 4.0: A supply chain innovation perspective. *International Journal of Production Research*, *58*(5), 1425–1441.

Horváth, D., & Szabó, R. Z. (2019). Driving forces and barriers of industry 4.0: Do multinational and small and medium-sized companies have equal opportunities? *Technological Forecasting and Social Change*. Elsevier, 146(October 2018), 119–132. doi: 10.1016/j.techfore.2019.05.021.

Kagermann, H., Wahlster, W., & Helbig, J. (2013). *Recommendations for implementing the strategic initiative INDUSTRIE 4.0*, in: Final report of the Industrie 4.0 Working Group, Acatech, Frankfurt am Main, Germany.

Lasi, H., Fettke, P., Kemper, H. G., Feld, T., & Hoffmann, M. (2014). Industrie 4.0. *Business & Information Systems Engineering*, *56*(4), 261–264.

Masood, T., & Sonntag, P. (2020). Industry 4.0: Adoption challenges and benefits for SMEs. *Computers in Industry*, *121*, 103261.

Müller, J. M., Buliga, O., & Voigt, K. I. (2018). Fortune favors the prepared: How SMEs approach business model innovations in Industry 4.0. *Technological Forecasting and Social Change*, *132*, 2–17.

Müller, J. M., Veile, J. W., & Voigt, K. I. (2020). Prerequisites and incentives for digital information sharing in industry 4.0 – An international comparison across data types. *Computers & Industrial Engineering*, *148*, 106733.

Moeuf, A., Lamouri, S., Pellerin, R., Tamayo-Giraldo, S., Tobon-Valencia, E., & Eburdy, R. (2020). Identification of critical success factors, risks and opportunities of industry 4.0 in SMEs. *International Journal of Production Research*, *58*(5), 1384–1400.

Sahi, G. K., Gupta, M. C., & Cheng, T. C. E. (2020). The effects of strategic orientation on operational ambidexterity: A study of Indian SMEs in the industry 4.0 era. *International Journal of Production Economics*, *220*, 107395.

Schmidt, M. C., Veile, J. W., Müller, J. M., & Voigt, K. I. (2020). Ecosystems 4.0: Redesigning global value chains. *The International Journal of Logistics Management*, *ahead-of-print*(ahead-of-print). https://doi.org/10.1108/IJLM-03-2020-0145

Stentoft, J., Adsbøll Wickstrøm, K., Philipsen, K., & Haug, A. (2020). Drivers and barriers for Industry 4.0 readiness and practice: Empirical evidence from small and medium-sized manufacturers. *Production Planning & Control*, 1–18.

Veile, J. W., Schmidt, M. C., Müller, J. M., & Voigt, K. I. (2020). Relationship follows technology! How industry 4.0 reshapes future buyer-supplier relationships. *Journal of Manufacturing Technology Management*, *ahead-of-print*(ahead-of-print). https://doi.org/10.1108/JMTM-09-2019-0318

2 Industry 4.0 towards a Circular Economy
Why Small and Medium-Sized Enterprises Must Be Integrated Better within Supply Chains

Petra Unterberger and Julian M. Müller

CONTENTS

2.1 INTRODUCTION

Industry 4.0 is driven by technologies such as the Internet of Things or cyber-physical systems that connect the physical and digital world. The concept of Industry 4.0 was introduced in 2011 at the Hanover Fair and received large interest from academia and practice since then (Kagermann et al., 2013). To exploit the full potential of Industry 4.0, relating to the concept of a fourth Industrial Revolution, holistic systems across entire value chains and lifecycles of products are necessary that enables smooth representations between these two worlds despite low transaction costs. The establishment of such holistic systems is a key requirement to extend the Linear Economy to a Circular Economy (Rizos et al., 2016; Zamfir et al., 2017) "that replaces the 'end-of-life' concept with reducing, alternatively reusing, recycling and recovering materials in production/distribution and consumption processes" (Kirchherr et al., 2017).

Based on the expected benefits of the fourth Industrial Revolution, many approaches have been developed that relate, so far, mostly to the application in "smart factories," but not across entire value chains (Bag et al., 2020; Birkel and Müller, 2021; de Sousa

DOI: 10.1201/9781003165880-2

Jabbour et al., 2018). However, the understanding of Industry 4.0 across entire supply chains, functional departments, or the entire lifecycle of products is required to approach the benefits of a circular economy but is significantly less developed than within single factories or firms. This can be attributed to several challenges for the implementation of these three forms of integration towards Industry 4.0, highlighted by extant research (Bag et al., 2020; Birkel and Müller, 2021; de Sousa Jabbour et al., 2018; Rahman et al., 2020; Müller et al., 2020). Especially downstream logistics and recycling activities are less understood from the perspective of Industry 4.0 (Bag et al., 2020; Rahman et al., 2020; Veile et al., 2020) and data continuity is a serious problem in practice. Small and medium-sized enterprises (SMEs) play a key role in the successful implementation of Industry 4.0 and the circular economy, as they have specific expertise and often niche-specific know-how that is of high importance for supply chains. However, SMEs face size-specific challenges, as feared transparency and lower digitization levels compared to larger enterprises (Kumar et al., 2020; Müller et al., 2020; Virmani et al.,2020). Therefore, the integration of SMEs is vital and these problems must be overcome, in order to enable a circular economy through Industry 4.0.

In response, this paper aims to shed light on challenges concerning data transparency for SMEs that hamper Industry 4.0 implementation towards a circular economy. Therefore, the concepts of Industry 4.0 and circular economy as well as specific characteristics of SMEs are outlined within the theoretical background. Then, existing research is used to specify the research gap concerning data transparency for SMEs in the context of Industry 4.0 towards a circular economy. Based on a qualitative literature analysis the challenges of interest are derived and further discussed. Finally, a conclusion and future research recommendations are given.

The interconnection of the physical and virtual world is essential for the establishment of Industry 4.0 and further for a circular economy. However, a successful implementation depends on the integration of SMEs, because of their high relevance within supply chains.

2.2 INDUSTRY 4.0

Industry 4.0 describes a concept initiated by the German government, indicating a fourth Industrial Revolution (Kagermann et al., 2013). An industrial revolution is characterized by fundamental changes for industrial value creation, but also for society and environment. Cyber-physical systems and the Internet of Things represent the technological foundation for Industry 4.0, which allow three forms of integration through digital technologies: Horizontal interconnection across the supply chain, vertical interconnection across functional departments, and end-to-end engineering, from product development to recycling, i.e., along the entire lifecycle of products (Kagermann et al., 2013). The concept of Industry 4.0 aims to gather, transmit, and analyze data throughout these three forms of integration, enabling several benefits.

2.3 CIRCULAR ECONOMY

In order to unlock the full potentials of Industry 4.0, these three forms of integration are necessary. This is especially true for the concept of the circular economy, which requires data consistency across both, entire supply chains, and entire lifecycles of products (Bag et al., 2020; de Sousa Jabbour et al., 2018; Rahman et al., 2020; Müller et al., 2020).

While a linear economy is limited to useful applications of materials by recycling or recovering, a circular economy is characterized by smarter product use and manufacture based on refusing, rethinking and reducing materials along the whole supply chain (Potting et al. 2017). New business models are essential for the establishment of a circular economy "with the aim to accomplish sustainable development, thus simultaneously creating environmental quality, economic prosperity and social equity, to the benefit of current and future generations" (Kirchherr et al., 2017). The need for new business models also arises when striving to exploit the potential of Industry 4.0, also in the context of SMEs (Müller and Voigt, 2018). This requirement is based on the fundamental reorientation of the circular economy, driven by the fourth Industrial Revolution towards integral systems. Therefore, both Industry 4.0 and the circular economy present several compound effects that must be explored further.

2.4 SMALL AND MEDIUM-SIZED ENTERPRISES

Raw materials processing activities and component manufacturing, as well as recycling activities, are often conducted by SMEs (Müller and Voigt, 2018). Hence, a better understanding the requirements of Industry 4.0 in from the perspective of SMEs is required in order to be able to contribute towards a circular economy.

SMEs, often defined as firms with up to 250 employees, have several characteristics that make it hard for them to approach the potentials of Industry 4.0. For instance, they possess limited resources and skills, they often acting as suppliers without direct end-customer contact and SMEs have low bargaining power compared to larger enterprises or lower digitization levels. Further, feared transparency of SMEs towards competitors, customers or third parties impedes the implementation of Industry 4.0 within entire supply chains (Kumar et al., 2020; Müller and Voigt, 2018; Virmani et al., 2020). However, to achieve end-to-end engineering, horizontal and vertical interconnection of the integration of SMEs is essential, as well as for data consistency, since they encompass 99.5 percent of all European companies and generate more than half of the net value added (European Commission, 2014).

2.5 LITERATURE REVIEW

In the following section, to set the stage for new research, prior knowledge about the relationships between Industry 4.0, the circular economy, and SMEs is reviewed. The great influence of the fourth Industrial Revolution on all industries supports extensive research in this regard. The need for a sustainable economy also involves a lot of research concerning circular economy.

The link between Industry 4.0 and the circular economy theoretically has been empirically discussed by several authors, but not specifically related to SMEs (Nascimento et al., 2019; Bag and Pretorius, 2020). For instance, Tseng et al. (2018) highlight the big data-driven industrial symbiosis through Industry 4.0 that drives social, environmental, and economic rethinking. However, one of their key findings is the limited implementation of data-driven recycle, reduce, and reuse approaches in practice. As another example, Rajput and Singh (2019) argue that connection of circular economy and Industry 4.0 is strongly driven by artificial intelligence, service, and an appropriate policy framework, as well as by the circular economy, while it is mostly challenged by the usage of interface designing and the application of an automated synergy model. Still, while those factors take a holistic view on the topic of Industry 4.0 and the concept of the circular economy, specific aspects of SMEs have not been explored in detail in extant research.

As with Industry 4.0 and the concept of the circular economy, there is prior knowledge that focuses on SMEs in the context of Industry 4.0. It is noticeable that the usage of Industry 4.0 within SMEs is often limited to a cost-driven perspective, including the adoption of cloud computing, the Internet of Things, and the monitoring of industrial processes, without the consideration of adequate business model transformations (Moeuf et al., 2018). However, to exploit the full potential of Industry 4.0, SME business models have to innovate their strategy concerning value creation, value capture, and value offer so that the benefits of new technologies are not restricted to single applications (Müller and Voigt, 2018). To gain a better understanding of the current Industry 4.0 development phase of SMEs, maturity models are needed that adequately reflect the specificities of SMEs and Industry 4.0 (Mittal et al., 2018).

As stated before, less research is available about SMEs' efforts towards a circular economy. Nevertheless, Zamfir et al. (2017) offer analysis of business strategies of SMEs concerning the establishment of a circular economy at the company level in reference to their characteristics and investment decisions. Further, Rizos et al. (2016) investigate, that approximately two-thirds of SMEs recognize the company environmental culture as the enabler of a circular economy. In contrast, lack of support supply and demand network, of capital, of government support, technical know-how, information, and administrative burden are critical barriers to establishing a circular economy for SMEs.

This literature overview shows that regardless of the analyzed relationship, SMEs' strategic limitation of outdated business models and operational constraints due to lack of data consistency along the entire supply chain represent the major barriers to establishing a circular economy via Industry 4.0. Further, Industry 4.0 as an enabler of a circular economy with specialization in the integration of SMEs has not yet been analyzed in detail.

In response and with focus on the literature gap this chapter aims to investigate challenges to data transparency for SMEs that are in conflict with Industry 4.0 implementation towards a circular economy. Therefore, a more specific qualitative literature analysis is applied.

2.6 CHALLENGES OF INDUSTRY 4.0 TOWARDS A CIRCULAR ECONOMY REGARDING SMEs

In extant literature, aspects such as a reduction of resource consumption and waste generation are named as potential benefits of Industry 4.0 towards a circular economy.

Examples include early recognition of errors, self-optimizing, and interconnected production and logistics processes, e.g., more efficient transport routes reducing resource consumption or predictive capabilities preventing waste production. Increased resource efficiency derives from better load and energy balancing across the supply chain, also allowing for the inclusion of renewable energy sources as their availability and energy demand can be better aligned. Another example includes products that last longer due to predictive capabilities. Further, data flowing back from product usage can be used to optimize product development or recycling, such as information required for upcycling or recycling processes based on data collected in product usage. Moreover, certification, monitoring, and traceability can be achieved more easily through transparency across the supply chain (Bag et al., 2020; Birkel and Müller, 2021; de Sousa Jabbour et al., 2018; Rahman et al., 2020; Müller et al., 2020).

In order to approach the potentials named earlier, data transparency across the supply chain is required. To gain data transparency, four development steps are necessary. First, the individual companies have to build their data infrastructure, which has to be rich in terms of quantity – for example due to usage of RFID and sensors as well as in terms of quality, so that irrelevant data is filtered out while valuable data is prepared. Second, data continuity across the entire supply chain needs to be established by data sharing through standardized data governing with predefined access and security regulations. Third, to create value out of the shared data, knowledge needs to be deduced, so that additionally analytical competences, especially regarding big data are required. Fourth, there must be a data-driven culture, where decision makers possess statistical understanding and critical thinking ability is essential to institutionalize the importance of data transparency (Arunachalam et al., 2018).

Data transparency, hence, requires SMEs to share and receive data, which rises several challenges from the perspective of SMEs (Müller and Voigt, 2018). Table 2.1

TABLE 2.1
Challenges of Industry 4.0 towards Circular Economy in SMEs

Category	Description of challenges towards data transparency
SMEs do not have access to data required	SME access is limited to direct data shared with their own customers and suppliers
	No access for SMEs to overall data generation in the supply chain
	SMEs that act as supplier do not have direct contact to the end customer
	No access for SMEs to product data in usage
SMEs make no efforts required for data sharing	Many different data standards and interfaces within SMEs
	Semi-automated and only partially digitized approaches within SMEs
	Strong efforts for digitization and standardization for SMEs
	SMEs cannot afford adequate solutions or specialists for data security
SMEs are reluctant to share data	SMEs fear data transparency
	Information could be used against SMEs in terms of replacement or price reduction
	Data transparency will increase the bargaining power of larger customers disproportional in their favor
	SMEs are afraid of data leaks to competitors or third parties

lists several challenges for SMEs derived from the literature that are in conflict with Industry 4.0 implementation towards a circular economy (Bag et al., 2020; de Sousa Jabbour et al., 2018; Kumar et al., 2020; Rahman et al., 2020; Mittal et al., 2018; Moeuf et al., 2018; Schmidt et al., 2020; Veile et al., 2020; Virmani et al., 2020).

2.7 DISCUSSION AND CONCLUSION

Many challenges derive from SMEs that are unable or reluctant to share data with their customers. This relates to technical considerations, strong efforts, and fears about data transparency. Further, SMEs do not receive data they require for optimizing their own product development, production, or recycling activities. This is because many SMEs often to not have direct contact to end customers but act as suppliers.

While economic benefits can also be generated from peer-to-peer data sharing within supply chains, ecological potentials towards a circular economy typically require data transparency across the supply chain. Since SMEs are often active in raw materials processing, component manufacturing, and recycling activities within supply chains, their integration is vital for developing a circular economy through Industry 4.0. Therefore, better integrating SMEs for data transparency is a requirement for policy and practice alike. Data transparency must be ensured across the supply chain, including SMEs receiving and sharing data. Especially in order to develop circular economy, it is vital to integrate SMEs in both directions: data sharing by SMEs and SMEs receiving data. Technical considerations, implementation efforts, and fears must be better understood in a holistic view, and countermeasures must be taken to better integrate SMEs within the concepts of Industry 4.0 and the circular economy.

In conclusion, this chapter investigates the literature to assess the challenges towards data transparency for SMEs that are in conflict with Industry 4.0 implementation towards a circular economy. Thereby, the different challenges bear on the reluctance of SMEs to share data or make efforts required for sharing data and on the lack of access to data required by SMEs.

The chapter subsumes the findings of extant literature, which only partially mentioned aspects of Industry 4.0 and the circular economy for SMEs, and highlights and condenses the most relevant aspects for SMEs for their integration within industrial value chains (Bag et al., 2020; de Sousa Jabbour et al., 2018; Kumar et al., 2020; Rahman et al., 2020; Mittal et al., 2018; Moeuf et al., 2018; Schmidt et al., 2020; Veile et al., 2020; Virmani et al., 2020).

This chapter is limited by not being based on our own empirical data or case studies and only condensing extant knowledge with a new focus. In the future, this study shall be extended to incorporate such an empirical investigation, focusing on aspects of multi-tier supply-chain management and relationships between large enterprises and SMEs.

Additional avenues for future research should involve a better understanding of entire supply chains. The interplay of technical considerations, implementation efforts, and fears regarding the data transparency of SMEs must be better understood, highlighting possible strategies in response. For instance, strategies could include decentralized and secure data storage facilities for end-to-end supply chains, affordable solutions for SMEs to share and receive data, and supporting SMEs directly in providing data, as their efforts are vital to develop Industry 4.0 potentials towards a circular economy. Further, aspects such as sustainable business models in the context of Industry 4.0 and circular economy must be better understood in the future.

REFERENCES

Arunachalam, D., Kumar, N., & Kawalek, J. P. (2018). Understanding big data analytics capabilities in supply chain management: Unravelling the issues, challenges and implications for practice. *Transportation Research Part E: Logistics and Transportation Review*, 114, 416–436.

Bag, S., & Pretorius, J. H. C. (2020). Relationships between industry 4.0, sustainable manufacturing and circular economy: Proposal of a research framework. *International Journal of Organizational Analysis*, in press, available online.

Bag, S., Wood, L. C., Mangla, S. K., & Luthra, S. (2020). Procurement 4.0 and its implications on business process performance in a circular economy. *Resources, Conservation and Recycling*, 152, 104502.

Birkel, H. S., & Müller, J. M. (2021). Potentials of industry 4.0 for supply chain management within the triple bottom line of sustainability – a systematic literature review. *Journal of Cleaner Production*, 125612.

de Sousa Jabbour, A. B. L., Jabbour, C. J. C., Godinho Filho, M., & Roubaud, D. (2018). Industry 4.0 and the circular economy: A proposed research agenda and original roadmap for sustainable operations. *Annals of Operations Research*, 270(1–2), 273–286.

European Commission. (2014). Evaluation of the SME definition. Final report. https://ec.europa.eu /growth/smes/sme-definition_en.

Kagermann, H., Wahlster, W., & Helbig, J. (2013). Recommendations for implementing the strategic initiative Industrie 4.0 – Final report of the Industrie 4.0 Working Group. Communication Promoters Group of the Industry-Science Research.

Kirchherr, J., Reike, D., & Hekkert, M. (2017). Conceptualizing the circular economy: An analysis of 114 definitions. *Resources, Conservation and Recycling*, 127, 221–232.

Kumar, R., Singh, R. K., & Dwivedi, Y. K. (2020). Application of industry 4.0 technologies in Indian SMEs for sustainable growth: Analysis of challenges. *Journal of Cleaner Production*, in press, available online.

Mittal, S., Khan, M. A., Romero, D., & Wuest, T. (2018). A critical review of smart manufacturing and Industry 4.0 maturity models: Implications for small and medium-sized enterprises (SMEs). *Journal of Manufacturing Systems*, 49, 194–214.

Moeuf, A., Pellerin, R., Lamouri, S., Tamayo-Giraldo, S., & Barbaray, R. (2018). The industrial management of SMEs in the era of Industry 4.0. *International Journal of Production Research*, 56(3), 1118–1136.

Müller, J. M., Veile, J. W., & Voigt, K. I. (2020). Prerequisites and incentives for digital information sharing in industry 4.0 – An international comparison across data types. *Computers & Industrial Engineering*, 148, 106733.

Müller, J. M., & Voigt, K. I. (2018). Sustainable industrial value creation in SMEs: A comparison between industry 4.0 and made in China 2025. *International Journal of Precision Engineering and Manufacturing-Green Technology*, 5(5), 659–670.

Nascimento, D. L. M., Alencastro, V., Quelhas, O. L. G., Caiado, R. G. G., Garza-Reyes, J. A., Rocha-Lona, L., & Tortorella, G. (2019). Exploring industry 4.0 technologies to enable circular economy practices in a manufacturing context. *Journal of Manufacturing Technology Management*, 30(3), 607–627.

Potting, J., Hekkert, M. P., Worrell, E., & Hanemaaijer, A. (2017). *Circular Economy: Measuring Innovation in the Product Chain* (No. 2544). The Hague: PBL Publishers.

Rahman, S. M., Perry, N., Müller, J. M., Kim, J., & Laratte, B. (2020). End-of-life in industry 4.0: Ignored as before? *Resources, Conservation and Recycling*, 154, 104539.

Rajput, S., & Singh, S. P. (2019). Connecting circular economy and industry 4.0. *International Journal of Information Management*, 49, 98–113.

Rizos, V., Behrens, A., Van der Gaast, W., Hofman, E., Ioannou, A., Kafyeke, T., Flamos, A., Rinaldi, R., Papadelis, S., Hirschnitz-Gabers, M., & Topi, C. (2016). Implementation of circular economy business models by small and medium-sized enterprises (SMEs): Barriers and enablers. *Sustainability*, 8(11), 1212.

Schmidt, M. C., Veile, J. W., Müller, J. M., & Voigt, K. I. (2020). Ecosystems 4.0: Redesigning global value chains. *The International Journal of Logistics Management*, ahead-of-print(ahead-of-print). https://doi.org/10.1108/IJLM-03-2020-0145

Tseng, M. L., Tan, R. R., Chiu, A. S., Chien, C. F., & Kuo, T. C. (2018). Circular economy meets industry 4.0: Can big data drive industrial symbiosis? *Resources, Conservation and Recycling*, 131, 146–147.

Veile, J. W., Schmidt, M. C., Müller, J. M., & Voigt, K. I. (2020). Relationship follows technology! How industry 4.0 reshapes future buyer-supplier relationships. *Journal of Manufacturing Technology Management*, ahead-of-print(ahead-of-print). https://doi.org/10.1108/JMTM-09-2019-0318

Virmani, N., Bera, S., & Kumar, R. (2020). Identification and testing of barriers to sustainable manufacturing in the automobile industry: A focus on Indian MSMEs. *Benchmarking: An International Journal*, in press, available online.

Zamfir, A. M., Mocanu, C., & Grigorescu, A. (2017). Circular economy and decision models among European SMEs. *Sustainability*, 9(9), 1507.

3 Industry 4.0 in Finland

Towards Twin Transition

Iqra Sadaf Khan, Osmo Kauppila, Jukka Majava, Marko Jurmu, Jan Olaf Blech, Elina Annanperä, Marko Jurvansuu, and Susanna Pirttikangas

CONTENTS

3.1 INTRODUCTION AND COUNTRY BACKGROUND

As the European Union's (EU) leader in digitalization performance (European Commission 2019, 2020) and its most sparsely populated nation, Finland is a Nordic country of 5.5 million inhabitants, with few political risks, high-end infrastructure, and good quality logistics (Kaivo-Oja *et al.* 2018). Finland is among the leaders in many global rankings, such as PISA (Schleicher 2019), anti-corruption (Transparency International 2020), happiness (Helliwell *et al.* 2020), and innovation (Cornell University *et al.* 2020).

In Industry 4.0 rankings, Finland also ranks near the top. Castelo-Branco *et al.* (2019) place Finland as a leader alongside the Netherlands in Industry 4.0 readiness in manufacturing in the EU. Atik and Ünlü (2019) rank Finland second in Europe in Industry 4.0 performance. Additionally, Sung (2018) investigated global Industry 4.0 competitiveness based on UBS (2016), WEF (2016), and IMD (2017) and ranked Finland second behind Singapore.

The Finnish industry is based on high-value-added export-oriented manufacturing (Ciffolilli and Muscio 2018) due to its small domestic market and price competition not being an option. To remain competent, Finnish manufacturers need to be flexible, reliable, and able to provide state-of-the-art technology (Kaivo-Oja *et al.* 2018). Examples of export companies in Finland include Stora Enso (wood and paper products), Kemira (chemicals), Wartsila (marine, power), Neste (oil products),

DOI: 10.1201/9781003165880-3

Nokia (information and communications technology), KONE (escalators), SSAB (steel products), ABB (robotics, power), and Ponsse (forest harvesters). Recently, the Finnish game industry has had success stories as well, such as Rovio (Angry Birds) and SuperCell (Clash of Clans).

Regarding exports, in 2005, nearly 86 percent of Finnish exports were industrial goods, but the share of services is currently 33 percent (Confederation of Finnish Industries 2020). Still, most exports are physical products. In 2019, the total goods exported were worth 64.8 billion euros with 2 percent annual growth, while the service exports totaled 31.7 billion euros in 2019 with 17 percent annual growth (Confederation of Finnish Industries 2020). The Finnish industry depends on its international supplier base, with more than 80 percent of intermediate goods and components coming from abroad (Ali-Yrkkö and Kuusi 2020).

The Finnish economy has fluctuated between periods of growth and turmoil due to events such as the collapse of Eastern trade accompanied by a financial crisis (1990–1993), a global financial crisis (2007–2009), and the collapse of Nokia's production (2007–2011). The shift towards digitalization could be seen to have started in the mid-1990s, when the value added in the electronics and electrical products sector exceeded both the wood and paper and machinery and equipment sectors. In recent years, between 2015 and 2020, the GDP has steadily grown (Statistics Finland), and there is also evidence of manufacturing jobs being reshored to Finland lately (Kaivo-Oja *et al.* 2018).

3.2 DIGITALIZATION POLICY IN FINLAND

Digitalization and Industry 4.0 under the term "Industrial Internet" can be found in policy documents from the era of the Katainen and Stubb governments (2011–2015). In Finland, strategizing preset visions and plans for the Industry 4.0 transition was the first step on the roadmap of Industry 4.0. For example, the governmental program of 2011 proffered the introduction of intelligent solutions in all sectors of society and the creation of intelligent strategies for each of the ministries (Prime Minister's Office 2011). In the *Industrial Competitiveness Approach* (Känkänen *et al.* 2013), information and communications technology (ICT) is recognized as not only a support function but a "bloodline" for the manufacturing industry.

Further development by the Ministry of Employment and the Economy ([MEE] 2014) has involved creating the prerequisites and support for the Industrial Internet's implementation. The report presented actions for promoting digitalization as a value creator of industrial manufacturing for the MEE itself, for the Finnish Funding Agency for Technology and Innovation (Tekes – an activator and funder of business, higher education, and research institutions' R&D projects), for the Ministry of Transport and Communications and for the Ministry of Education. Following this trend, in 2014, the Prime Minister's Office commissioned an assessment on the challenges and opportunities of the "Finnish Industrial Internet" (Ailisto *et al.* 2015). The results were reported in 2015, and another set of recommendations were suggested.

The Sipilä government's (2015–2019) program (Prime Minister's Office 2015) presented a new strategy that focused on the digitalization of public services and the creation of a growth environment. This led to the second phase on the Industry 4.0 roadmap, which emphasized the adoption of new technologies and solutions development by fostering a culture of experimentation. One action point of the program was to establish a governmental program for the Internet of Things. Later on, in 2017, AI was promoted to a strategic spearhead, and a roadmap for integrating AI into the Finnish society and businesses was created (MEE 2019). Ecosystem thinking also became visible in the national policy and was prominently featured in the governmental action plan for 2018–2019 (Finnish Government 2018). This can be viewed as the third step of Industry 4.0 implementation in Finland – ecosystemic collaboration and data-driven value creation.

The Rinne (2019) and the current Marin (2019–) governmental programs (Finnish Government 2019a, 2019b) advocated for a switch to sustainability, inclusiveness and carbon-neutrality. Ecosystemic thinking has also been strongly featured (e.g., "Ecosystems will be the engines of sustainable growth" [Finnish Government 2019a, 2019b]). The Research, Development, and Innovation Roadmap (Finnish Government 2020) lists three strategic development targets: competence, a new partnership model and an innovative public sector. A key figure in operationalizing the strategy is Business Finland (former Tekes), and their programs and terminology provide insight on how Industry 4.0 has been interpreted in Finland. Their policy is also transitioning towards sustainability through "Sustainable Manufacturing Finland" (Mattila 2020). This program shares the "twin transition" ideology of the EU's Green Deal, and implies promoting digitalization – especially through a platform economy, artificial intelligence (AI), and data-driven business models – while simultaneously increasing the sustainability of the business in terms of a circular economy, lowered carbon emissions and inclusion. This fourth step of the progression can be called "Sustainable Industry 4.0 Ecosystems."

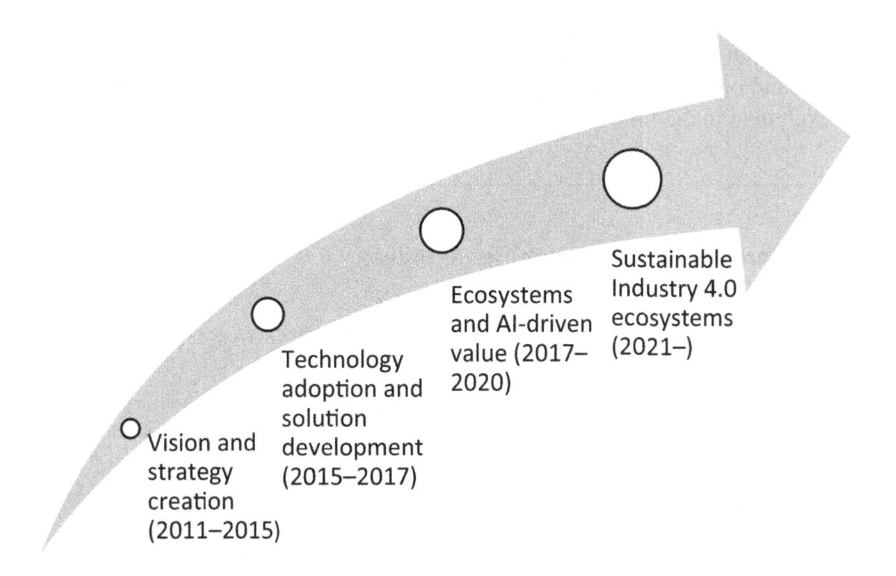

FIGURE 3.1 Evolution of the Industry 4.0 strategy in Finland.

It should be noted that despite Finland being on the forefront of digitalization, there has not been a national definition for Industry 4.0. "Industrial Internet" made a somewhat brief appearance in the mid-2010s, but it was soon superseded by digitalization and individual areas of Industry 4.0, such as the Internet of Things (IoT) and AI. Following, Kaivo-Oja *et al.* (2018) defined that Industry 4.0 "is marked by highly developed automation and digitization processes and by the use of electronics and information technologies (IT) in manufacturing and services." To summarize how this has taken place, Figure 3.1 illustrates key developments in the national Industry 4.0 strategy in Finland.

3.3 METHODOLOGY

To conduct more thorough investigations of drivers, barriers, and future opportunities for Finnish Industry 4.0 implementation, a framework of these three dimensions was first created based on a literature review of the subject. The most relevant literature sources identified include Türkes *et al.* (2019), Stentoft *et al.* (2019), Rajput and Singh (2019), Moeuf *et al.* (2020), Müller and Voigt (2018), Kamble *et al.* (2018), and Horváth and Szabó (2019). An initial search was constructed using the search strings "Industry 4.0 in Finland" and "Drivers, barriers and opportunities of Industry 4.0" in Google Scholar. After that, the best papers and reports specifically relevant to the Finnish context were included in the study. The contents of the papers and reports were carefully reviewed, and Industry 4.0-based barriers, drivers, and opportunities were handpicked using the identified documents.

Subsequently, each individual item's relevance to Finland was analyzed based on existing policies and research. This was complemented by the authors' expertise on the topic that was gained, for example, from conducting several research projects, such as Reboot IoT Factory (2018–2021). The project is one of the first pilot projects in the Finnish digitalization program Reboot IoT Finland, which aims to facilitate the digital transformation of Finnish manufacturing. Table 3.1 presents the drivers, barriers, and opportunities in order by their perceived relevance to the Finnish context, explained in the following sections.

TABLE 3.1
Drivers, Barriers, and Opportunities of Industry 4.0 in Finland

Drivers	Barriers	Opportunities
High market competition	Single source solutions for unique needs are difficult to find	Growing global markets
National infrastructure that supports Industry 4.0	SMEs lacking monetary and strategic support	Competitive advantage through smart and sustainable manufacturing
Culture of innovation and R&D	SMEs hesitance to digitalize their operations	Involving SMEs to create new innovations in business ecosystems
Highly skilled workforce	Aging population, less potential employees with new skills	Societies as innovation platforms

Drivers	Barriers	Opportunities
Culture of triple helix collaboration	Issues related to data security, ownership, and trust in digital ecosystems	Collaboration platforms for industrial symbiosis, value creation networks and business ecosystems
Business model innovation	Lack of methodological approaches, such as best-practice-examples, toolsets and distilled information	Sustainable business models, organizations, processes, and products

3.4 INDUSTRY 4.0 DRIVERS IN FINLAND

Changes in economic and social life conditions have led to reliance on modern digital technologies signifying high-tech strategies and innovations as the underlying factors of Industry 4.0 and its development (Türkes *et al.* 2019). The Finnish manufacturing industry has initiated digitalization processes to increase productivity, which is required for competency in the global market and retaining the industry in Finland. The industry has become a part of the "new globalization" trend the world has been facing since 1990 (Ali-Yrkkö *et al.* 2017).

Currently, Finnish factories are in very close competition with their global counterparts. High labor and logistics costs have resulted in decreased cost competitiveness. The total logistics costs of Finnish manufacturing firms are approximately 14 percent of the turnover, whereas in Swiss manufacturing firms, they are approximately 8 percent. The difference is partly attributable to the long transportation distance to export markets (Solakivi *et al.* 2018). To beat the global competition, the Finnish manufacturing industry relocated their activities and tasks to locations abroad to optimize their efficiency and the reconfiguration of their value-chains (Kaivo-Oja *et al.* 2018). Because the global markets are becoming more heterogenous over time, a growing number of competing companies opting for technological advancements have abruptly made it essential for Finnish companies to reduce their time-to-market rates, to be among the first movers and to gain a decisive advantage over their competitors by increasing their innovation capability, productivity, and efficiency.

Political support from different actors in Finland has played a vital role in shaping the Industry 4.0-based economic development in the information and telecommunications and healthcare and engineering sectors. With this support, Finland has been able to provide the necessary modern and robust infrastructure to support digitalization. Connectivity is targeted through immense 5G initiatives that aim to serve an extensive range of sectors, such as connected automobility, e-health and energy management. Furthermore, the infrastructure supports the use of advanced digital technologies, such as AI, the IoT, cyber-physical systems (CPS), and big data in general. More specifically, Finland has been marked as the most advanced country to uptake cloud services (European Commission 2020).

The general aim of Industry 4.0 is to eradicate the boundaries between the digital and the physical world, acquiring highly skilled Industry 4.0 human operators for the accurate exchange of information between intelligent support systems, supported by their robust aids and capabilities (Schmidt *et al.* 2015). In Finland, such production systems support the digitization of production units by successful human-robot collaboration, thereby building smart and intelligent factories with multiplied efficiency levels in terms of monitoring and supervision support systems, digital computing systems, virtual trainings, and decision support systems. A highly educated workforce/human capital in terms of digitally skilled labor and ICT specialists has supported Finland in becoming a digitalization leader (European Commission 2020). This has been further reinforced by the Finnish national core curriculum for basic education (Finnish National Agency for Education 2020), which includes coding and programming from the very beginning of students' school education. For example, sixth-grade students studying handicrafts learn to embed automation as a part of their product, whereas students studying mathematics learn to solve problems in a graphical programming environment. Furthermore, in secondary school (grades 7–9), they learn how to use their skills to produce digital work individually and in a collaborative environment.

Finland has a tradition of research – business collaboration, and around 70 percent of large businesses collaborate on innovation with higher education or research institutions (OECD 2013, 2017). While the percentage of large businesses collaborating with research and higher education institutions is the highest in the world, the SME portion is only ranked fourth, after Great Britain, Belgium, and Austria. Collaboration between companies and research organizations produces new Industry 4.0-based innovations and operators in the 'factory of the future' (Isabel *et al.* 2019). Successful collaborations have presented key innovations together with smart technologies based on CPS, the IoT, cloud computing, big data, and 3D printing, leading to increased efficiency and competitiveness.

Industry 4.0 has established digitalization trends that vary from country to country. While the digitalization policies of Germany and Japan have focused primarily on product quality (Türkes *et al.* 2019), the US and China have emphasized efficient product delivery and cutting the costs, respectively (Urciuoli *et al.* 2013; Müller and Daeschle 2018; Zhu and Geng 2013). Large businesses in Finland have developed their business models based on new product and service offerings and the simplification of smart products based on quality. Finnish manufacturers distinguish themselves based on the reliability and flexibility they offer in the regional balancing of their value chains (Kaivo-Oja *et al.* 2018). This allows Finnish manufacturers to synchronize their operations with all the stakeholders in the value-chain and to precisely be more responsive to customer needs. Additionally, strong customer orientation, flexibility, and state-of-the-art technologies have resulted in the dynamic capabilities needed for Industry 4.0 development (Xu *et al.* 2018). In summary, the high process digitalization of the country is based on high-end governmental financial support, an educated workforce, technological innovation and close industry – research collaborations.

3.5 INDUSTRY 4.0 BARRIERS IN FINLAND

The Finnish industry is characterized more by customized offerings, which makes it hard to find off-the-shelf/single-source solutions. To find holistic solutions, companies need the competence to define their own needs and collaborate with the SMEs to provide expanded solutions. Subsequently, without the large budget required for SMEs to match big corporations, it becomes a huge challenge for SMEs to achieve visibility, thereby making it difficult to be competitive and seen as valuable for strategic partnership.

Finland lacks the culture and resources for later-stage venture capital investments compared to many other countries (Saarikoski *et al.* 2014), which hinders its SMEs' growth in international markets. Furthermore, while the start-up culture itself is quite strong, there is a lack of start-ups oriented towards manufacturing industry innovation. In addition to venture capital, SMEs, particularly microenterprises, are underrepresented in government support for companies. For instance, micro and small enterprises only represent 6.2 percent and 22 percent, respectively, of subsidy receivers (Statistics Finland 2020). Additionally, regarding Industry 4.0, digitalization money might not be targeted towards manufacturing.

Digitalization requires investments in new technologies, such as artificial intelligence, digital twins, advanced robotics, and virtual reality. The field of new technologies is very wide and disperse, often without standards. However, even if the technology sounds lucrative, it is often hard to estimate the business benefits it can provide. Even large companies do not have enough resources to identify, evaluate, test, and pilot all digitalization solutions to gain their full benefits. Testing is required to achieve confidence in actual large-scale investments and deployment, and change management is required throughout the organization while processes become more autonomous. For SMEs and mid-cap size manufacturing, which comprise most Finnish exports, the challenge is even more severe. The profitable digitalization solutions identified by the forerunner companies would have to be scaled down to typically low-volume SME production. Additionally, the investment level needs to be lower for new technology in SMEs, as they may lack digital expertise.

Finland has experienced a natural decrease in population growth leading to a lack of young people for future the technological development (Santos *et al.* 2017), creating a gap between the skills of traditional factory workers and the new skills needed on the job. This is significant, as a Deloitte study has predicted that technology will likely create more jobs in the manufacturing industry (Wellener *et al.* 2020), posing a challenge for procuring skilled and motivated labor for the industry in Finland. Thus, as production costs are high in Finland and most likely will stay that way, the level of automation and optimization of production processes must be under constant development to keep the competitive edge.

Despite a well-established digital ecosystem, the country still needs substantial investment from companies in data security and protection standards. Further, the companies need to understand data security laws and standardization issues related to digital strategy and working with machines (Wang *et al.* 2015). These issues can be further tackled by offering new incentive initiatives and funding programs. Another shortcoming stems from the legal and contractual uncertainties in using certain technologies; however, these can be solved by developing and adopting legal frameworks about big data collection, data privacy and data security.

Finally, companies joining in implementing Industry 4.0 lack best-practice examples from successful organizations. Sharing methodological toolkits in the ecosystems will thereby create enormous opportunities for start-ups and SMEs in terms of providing a baseline for strategic and market orientations, exemplifying business operations (Sahi *et al.* 2020). Thus, while Finland is still striving to offer large-scale digitalization solutions in the manufacturing industry as a country with both strong ICT and manufacturing verticals, it could do much more.

3.6 INDUSTRY 4.0 OPPORTUNITIES IN FINLAND

Industry 4.0 linked with high-level digitalization has created the opportunity for Finnish companies to join global value chains and understand the diversity of different industrial branches, their economic geography and their supply chain implementation strategies (Ailisto *et al.* 2015). As a result, this kind of multidimensional global understanding will help Finnish companies to attract job markets and foreign direct investments. Finland's strengths as a pilot plant site due to its low collaboration barrier, the availability of technology and its highly skilled workforce should be further harnessed.

Industry 4.0 has and continues to transform the definition and skills required for workers. Analytical thinking and innovation, active learning, complex problem-solving, critical thinking and analysis, creativity, originality and initiative, leadership and social influence, technology uses monitoring and control, and technology designs are the top skills for 2025 according to the World Economic Forum (WEF 2020). The education curriculum and systems must evolve to support the lifelong learning of blue-collar workers and white-collar workers as well as to attract individuals to counter the labor shortage in the manufacturing industry. To ensure the rapid reskilling of society, agile education is required to ensure that companies can leverage novel solutions.

Previously, SMEs in Finland have lagged behind in adopting Industry 4.0 due to multiple factors, such as lack of resources compared to larger companies and a lower degree of initiative to apply the technologies within their business networks (Stentoft *et al.* 2019). However, SMEs and start-ups are now seen as tools for enhancing the strategic and operative performance of the traditional companies (Ailisto *et al.* 2015). Ecosystem collaboration projects and the growth of a start-up scene that fosters the creation of more Industry 4.0-oriented start-ups has resulted in companies such as Meluta (acoustic and vibroacoustic measurements) and Visual Components (3D manufacturing simulation). However, even though SMEs are involved in Industry 4.0 implementation as technology and solution providers, the level of SMEs' Industry 4.0 maturity in absorptive capacity and knowledge acquisitions (Müller *et al.* 2020) could be further improved.

The value created based on Industry 4.0 and digitalization in Finland has utilized smart components, sensors, data sharing standards, and interfaces building autonomous and integrative architectures. Currently, it is necessary to focus on the interconnected systems forming trusted and collaborative networks, thus creating a need for innovative societies. These platforms and ecosystems based on innovative societies are seizing new opportunities in building new strategic and operative business capabilities to integrate processes, structures, visions, information systems, data and competencies through active experience sharing. In addition, cross-company collaborations for exchanging smart data, resources, products and materials form value-creating networks, offering new opportunities for industrial symbiosis and thus creating ways for closed-loop product lifecycles through efficient coordination (Schuh *et al.* 2014). In an attempt to capture further value, Finland is currently in the stage of moving past ecosystemic collaboration into collaboration between ecosystems.

Another major global disruptor of late has been platform economy, with giants such as Google or Amazon aggressively utilizing the "winner takes all" dynamic to capture the B2C market in digital services. To address these developments, the EU has devised the European Data Strategy, which established a regulatory framework for handling and utilizing data in business value creation. It specifically assesses value capturing in the emerging B2B platform economy based on data sharing between companies, organizations, and the public sector. Implementations of the data strategy include the International Data Spaces (IDS) and GAIA-X architectures, which aim for a federated architecture of services for B2B data sharing and value creation. At the close of 2020, GAIA-X was also emerging in Finland. The Finnish hub is forming, and the first GAIA-X members have included VTT, SITRA, CSC, and the Vastuu Group. The potential for cross-vertical data sharing in the Industry 4.0 verticals has been recognized. However, use cases are still few, and opportunities are more unclear in contrast to some of the other verticals moving forward in the EU, such as mobility-as-a-service and healthcare. Nevertheless, Finland has the prerequisites in place and has profiled itself along with the Netherlands as one of the frontrunners in charting opportunities arising from GAIA-X (Vahti 2020).

Moreover, the technological advancements and inclinations based on Industry 4.0 require companies to transform from a linear to a circular economy and forge a path towards sustainability (Rajput and Singh 2019). Manufacturing companies in Finland are also striving to attain sustainable operations and achieve circular economy principles. Exploring the opportunities created by the technological revolution, companies are transforming their supply chains by generating a vast amount of data concerning raw materials, waste monitoring, energy consumption, closed loop supply chains, and assessments of real-time information (Geissdoerfer *et al.* 2017). In a nutshell, if companies rightfully use the opportunities provided by Industry 4.0, they can successfully develop sustainability-based business models, thereby maintaining a balance between economic, environmental, and social aspects (Rajput and Singh 2019).

3.7 DISCUSSION AND CONCLUSION

Throughout the 2010s, digitalization was recognized as a key factor of Finland's global competitiveness, and the elements of Industry 4.0 have been quite successfully implemented through consistent national policy. This has been supported by a high level of education and the development of a supporting national infrastructure (OECD 2017). The traditions of a collaborative and trust-based business environment, triple helix collaboration, and a strong culture of experimentation, rapid testing, and innovation have advanced Industry 4.0 implementation (Ailisto *et al.* 2015; Schuh *et al.* 2014). Moreover, high levels of R&D investments and a transition to innovation ecosystem-based thinking for value creation (Finnish Government 2018) helped Finland to reach its position as one of the global front-runners of Industry 4.0 maturity in 2019 (Atik and Ünlü 2019; Castelo-Branco *et al.* 2019).

However, areas of improvement and barriers to implementation have been recognized. Finland's population is one of the oldest in Europe and is rapidly aging (Finnish Institute for Health and Welfare 2020), causing serious concerns for recruiting a workforce with an up-to-date set of skills. While major companies are advancing in digitalization, the knowledge is not always disseminated throughout their supply networks. In the manufacturing sector and within SMEs, there is still room for the implementation of Industry 4.0 tools, particularly related to analytics, as noted by Mittal *et al.* (2020). Improving digital skills in SMEs is also a key objective in the national AI 4.0 program launched in late 2020 (MEE 2020).

The focus on high-value-adding niche markets indicates that implementing off-the-shelf solutions is often impossible, resulting in higher costs and lead times in the implementation of digital solutions. This contributes to another barrier of Industry 4.0 implementation, as costs are often assessed as high compared to the benefits. This holds particularly true for SMEs, and the barriers of lack of expertise as well as unwillingness to commit to a long-term Industry 4.0 strategy or to invest in technology that could soon be obsolete (Moeuf *et al.* 2020) can be recognized in Finnish SMEs as well.

Many future opportunities can be recognized regarding the current twin transition towards digitalization and sustainability. With the ecosystem thinking that has already been established, a current initiative to harness the knowledge and innovation created across vertical clusters aims to establish a national "supercluster" or an "ecosystem of ecosystems" to nationally coordinate these activities. This Sustainable Industry X (SIX) initiative (Figure 3.2) aims to integrate existing clusters and ecosystems, the best practices, national and EU policies, industry and stakeholder needs, and triple helix collaboration. It is currently in an early start-up phase, but if successful, it could boost national competitiveness and support the goals of sustainable manufacturing in the near future.

Other future opportunities can be found within the current ecosystem collaboration models. Many of these initiatives are project-based, and they operate based on public funding. The challenge of sustaining them after the funding period has passed has not been fully solved. Furthermore, finding more explicit value propositions from research to industry and vice versa could support the creation of partnerships spanning beyond public funding of ecosystem collaboration projects. Moreover, active information sharing and working together could establish a basis of trust and more sustainable collaboration. The research community is also responsible for finding these new management qualities and practices (Horváth and Szabó 2019).

This leads to another future challenge for increasing SME participation in these innovation ecosystems, both as solution providers and solution utilizers. Due to their limited resources and ability to take risks, SMEs need implementation support to exploit and to explore Industry 4.0 opportunities, as observed by Müller *et al.* (2020). Future opportunities also exist within the platform and data economies. Last, to

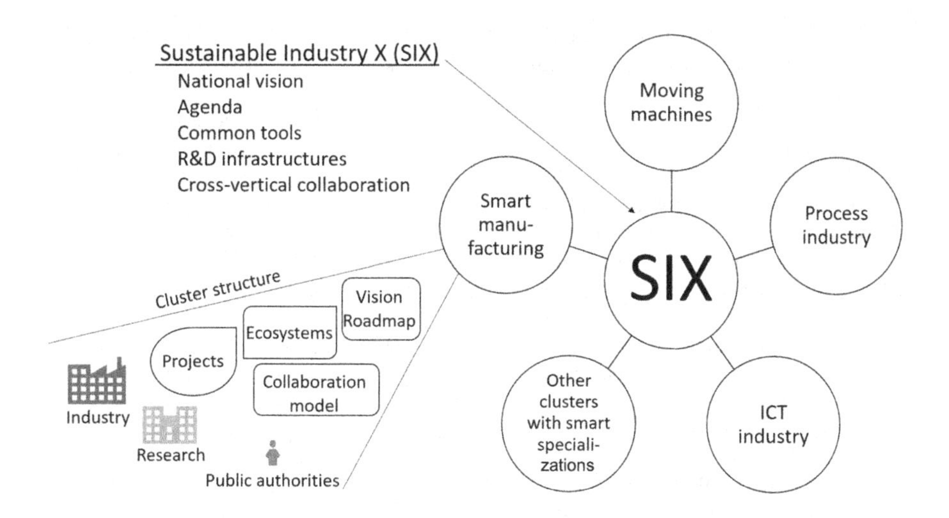

FIGURE 3.2 Initiative for establishing a national industry program SIX

counter the issues stemming from the aging population and the shifting required skillset, all parties of the triple helix have to find ways to further promote lifelong learning and to quickly adapt to the changing requirements of workforce competences (Isabel *et al.* 2019).

Overall, as a society, Finland has remained at the forefront of digitalization for some time now, which has also enabled Finnish industries to successfully adopt and develop Industry 4.0 solutions and new operating models. However, structural characteristics of the society, such as the aging population and high labor cost, indicate that keeping on top of global trends and technological change remains necessary for Finland to remain globally competitive. How to adapt and stay on the forefront of innovation remains a primary concern, as sustainability and circular-economy-based operating models are becoming a new standard in global business.

3.8 ACKNOWLEDGMENTS

This chapter is dedicated to the memory of one of the authors, Jan Blech, who unexpectedly passed away in February 2021.

REFERENCES

Ailisto, H., Mäntylä, M., Seppälä, T., Collin, J., Halén, M., Juhanko, J., Jurvansuu, M., Koivisto, R., Kortelainen H., Simons, M., Tuominen, M., and Uusitalo, T., 2015. Finland – the Silicon Valley of industrial internet. Publications of the Government's analysis, assessment and research activities.

Ali-Yrkkö, J., and Kuusi, T., 2020. Korona-sokki talouteen – Missä määrin Suomi on riippuvainen ulkomaisista arvoketjuista? [Corona-shock hits the economy – to what extent Finland is dependent on global value chains?]. *ETLA Muistio*, 87. Available from: https://pub.etla.fi/ETLA-Muistio-Brief-87.pdf [Accessed 09 December 2020].

Ali-Yrkkö, J., Lehmus, M., Rouvinen, P., and Vihriälä, V., 2017. *Riding the Wave: Finland in the Changing Tides of Globalisation*. Helsinki: Research Institute of the Finnish Economy – ETLA.

Atik, H., and Ünlü, F., 2019. The measurement of industry 4.0 performance through industry 4.0 index: An empirical investigation for Turkey and European countries. *Procedia Computer Science*, 158, 852–860. Available from: https://doi.org/10.1016/j.procs.2019.09.123 [Accessed 09 December 2020].

Castelo-Branco, I., Cruz-Jesus, F., and Oliveira, T., 2019. Assessing industry 4.0 readiness in manufacturing: Evidence for the European Union. *Computers in Industry*, 107, 22–32. Available from: https://doi.org/10.1016/j.compind.2019.01.007 [Accessed 09 December 2020].

Ciffolilli, A., and Muscio, A., 2018. Industry 4.0: National and regional comparative advantages in key enabling technologies. *European Planning Studies*, 26(12), 2323–2343. Available from: https://doi.org/10.1080/09654313.2018.1529145 [Accessed 09 December 2020].

Confederation of Finnish Industries, 2020. Ulkomaankauppa [Foreign Trade]. Available from: https://ek.fi/tutkittua-tietoa/tietoa-suomen-taloudesta/ulkomaankauppa/ [Accessed 24 November 2020].

Cornell University, INSEAD, and WIPO, 2020. *The Global Innovation Index 2020: Who Will Finance Innovation?* Ithaca, Fontainebleau, and Geneva: WIPO. Available from: www.wipo.int/edocs/pubdocs/en/wipo_pub_gii_2020.pdf [Accessed 09 December 2020].

European Commission, 2019. Digital Economy and Society Index (DESI) 2019. Available from: https://digital-strategy.ec.europa.eu/en/library/digital-economy-and-society-index-desi-2019 [Accessed 23 August 2021].

European Commission, 2020. The Digital Economy and Society Index (DESI). Available from: https://digital-strategy.ec.europa.eu/en/policies/desi [Accessed 23 August 2021].

Finnish Government, 2018. Finland, a land of solutions: Government action plan 2018–2019. *Publications of the Finnish Government*, 29.

Finnish Government, 2019a. Programme of Prime Minister Antti Rinne's government 6 June 2019: Inclusive and competent Finland – a socially, economically and ecologically sustainable society. *Publications of the Finnish Government*, 25.

Finnish Government, 2019b. Programme of Prime Minister Sanna Marin's government 6 June 2019: Inclusive and competent Finland – a socially, economically and ecologically sustainable society. *Publications of the Finnish Government*, 33, 228.

Finnish Government, 2020. The national roadmap for research, development and innovation. Available from: https://minedu.fi/en/rdi-roadmap [Accessed 09 December 2020].

Finnish Institute for Health and Welfare, 2020. Ageing policy. Available from https://thl.fi/en/web/ageing/ageing-policy [Accessed 03 December 2020].

Finnish National Agency for Education, 2020. *National Core Curriculum for Basic Education 2014*. 3rd edition. Helsinki: Finnish National Agency for Education.

Geissdoerfer, M., Savaget, P., Bocken, N. M., and Hultink, E. J., 2017. The circular economy: A new sustainability paradigm? *Journal of Cleaner Production*, 143, 757–768. Available from: https://doi.org/10.1016/j.jclepro.2016.12.048 [Accessed 09 December 2020].

Helliwell, J. F., Layard, R., Sachs, J., and De Neve, J-E., eds., 2020. *World Happiness Report 2020*. New York: Sustainable Development Solutions Network.

Horváth, D., and Szabó, R. Z., 2019. Driving forces and barriers of industry 4.0: Do multinational and small and medium-sized companies have equal opportunities? *Technological Forecasting and Social Change*, 146, 119–132. Available from: https://doi.org/10.1016/j.techfore.2019.05.021 [Accessed 09 December 2020].

IMD, 2017. *IMD World Digital Competitiveness Ranking*. Lausanne: IMD Switzerland.

Isabel, C., Cruz-Jesus, F., and Oliveira, T., 2019. Assessing industry 4.0 readiness in manufacturing: Evidence for the European Union. *Computers in Industry*, 107, 22–32. Available from: https://doi.org/10.1016/j.compind.2019.01.007 [Accessed 09 December 2020].

Kaivo-Oja, J., Knudsen, M. S., and Lauraeus, T., 2018. Reimagining Finland as a manufacturing base: The nearshoring potential of Finland in an industry 4.0 perspective. *Business Management and Education*, 16, 65–80. Available from: https://doi.org/10.3846/bme.2018.2480 [Accessed 09 December 2020].

Kamble, S. S., Gunasekaran, A., and Sharma, R., 2018. Analysis of the driving and dependence power of barriers to adopt industry 4.0 in Indian manufacturing industry. *Computers in Industry*, 101, 107–119. Available from: https://doi.org/10.1016/j.compind.2018.06.004 [Accessed 09 December 2020].

Känkänen, J., Lindroos, P., and Myllylä, M., 2013. Elinkeino- ja teollisuuspoliittinen linjaus – Suomen talouskasvun eväitä 2010-luvulla [Industrial Competitiveness Approach. Means to guarantee economic growth in Finland in the 2010s]. MEE Publications: Innovation.

Mattila, T., 2020. The pressures for renewal in the manufacturing industry are intensifying – is a "twin transition" possible? [Blog post]. Available from: www.businessfinland.fi/en/whats-new/blogs/2020/the-pressures-for-renewal-in-the-manufacturing-industry-are-intensifying [Accessed 09 December 2020].

Ministry of Economic Affairs and Employment, 2020. Artificial intelligence 4.0 programme to speed up digitalisation of business [Press release]. Available from: https://valtioneuvosto.fi/en/-/1410877/artificial-intelligence-4.0-programme-to-speed-up-digitalisation-of-business [Accessed 13 November 2020].

Ministry of Employment and the Economy, 2014. Teollisuus osana elinvoimaista elinkeinorakennetta. [Industry as a of a vital occupational structure. Global trends in industry, Finland's industrial situation and the the checkmarks for renewable Finnish Industry]. MEE Publications: Innovation.

Ministry of Employment and the Economy, 2019. Edelläkävijänä tekoälyaikaan: Tekoälyohjelman loppuraportti 2019. [Towards the AI age as a front runner: The final report of the AI programme 2019]. MEE Publications.

Mittal, S., Khan, M. A., Purohit, J. K., Menon, K., Romero, D., and Wuest, T., 2020. A smart manufacturing adoption framework for SMEs. *International Journal of Production Research*, 58(5), 1555–1573. Available from: https://doi.org/10.1080/00207543.2019.1661540 [Accessed 09 December 2020].

Moeuf, A., Lamouri, S., Pellerin, R., Tamayo-Giraldo, S., Tobon-Valencia, E., and Eburdy, R., 2020. Identification of critical success factors, risks and opportunities of industry 4.0 in SMEs. *International Journal of Production Research*, 58(5), 1384–1400. Available from: https://doi.org/10.1080/00207543.2019.1636323 [Accessed 09 December 2020].

Müller, J. M., Buliga, O., and Voigt, K. I., 2020. The role of absorptive capacity and innovation strategy in the design of industry 4.0 business Models-A comparison between SMEs and large enterprises. *European Management Journal*. Available from: https://doi.org/10.1016/j.emj.2020.01.002 [Accessed 09 December 2020].

Müller, J. M., and Daeschle, S., 2018. Business model innovation of industry 4.0 solution providers towards customer process innovation. *Processes*, 6, 260. Available from: https://doi.org/10.3390/pr6120260 [Accessed 09 December 2020].

Müller, J. M., and Voigt, K. I., 2018. Sustainable industrial value creation in SMEs: A comparison between industry 4.0 and made in China 2025. *International Journal of Precision Engineering and Manufacturing-Green Technology*, 5(5), 659–670. Available from: https://doi.org/10.1007/s40684-018-0056-z [Accessed 09 December 2020].

OECD, 2013. *Regions and Innovation: Collaborating Across Borders*. Paris: Organisation for Economic Cooperation and Development.

OECD, 2017. *OECD Science, Technology and Industry Scoreboard 2017: The Digital Transformation*. Paris: OECD Publishing. Available from: https://doi.org/10.1787/9789264268821-en [Accessed 09 December 2020].

Prime Minister's Office, 2011. *Programme of Prime Minister Jyrki Katainen's Government*. Helsinki: Ministry of Finance.

Prime Minister's Office, 2015. *Ratkaisujen Suomi: Pääministeri Juha Sipilän hallituksen strateginen ohjelma* [Finland of Solutions: a strategy programme of Prime Minister Juha Sipilä's government]. Helsinki: Prime Minister's Office.

Rajput, S., and Singh, S. P., 2019. Industry 4.0 – challenges to implement circular economy. *Benchmarking: An International Journal*. Available from: https://doi.org/10.1108/BIJ-12-2018-0430 [Accessed 09 December 2020].

Saarikoski, M., Roine, P., Ruohonen, J., Halonen, A., Sulin, J., and Lebret, H., 2014. *Evaluation of Finnish Industry Investment Ltd.* Helsinki: Ministry of Employment and the Economy.

Sahi, G. K., Gupta, M. C., and Cheng, T. C. E., 2020. The effects of strategic orientation on operational ambidexterity: A study of Indian SMEs in the industry 4.0 era. *International Journal of Production Economics*. Available from: https://doi.org/10.1016/j.ijpe.2019.05.014 [Accessed 09 December 2020].

Santos, C., Mehrsai, A., Barros, A. C., Araújo, M., and Ares, E., 2017. Towards industry 4.0: An overview of European strategic roadmaps. *Procedia Manufacturing*, 13, 972–979. Available from: https://doi.org/10.1016/j.promfg.2017.09.093 [Accessed 09 December 2020].

Schmidt, R., Möhring, M., Härting, R. C., Reichstein, C., Neumaier, P., and Jozinović, P., 2015. Industry 4.0-potentials for creating smart products: Empirical research results, June. In: *International Conference on Business Information Systems*, 16–27. Cham: Springer. Available from: https://doi.org/10.1007/978-3-319-19027-3_2 [Accessed 09 December 2020].

Schleicher, A., 2019. *PISA 2018: Insights and Interpretations*. Paris: OECD.

Schuh, G., Potente, T., Wesch-Potente, C., Weber, A. R., and Prote, J-P., 2014. Collaboration mechanisms to increase productivity in the context of Industry 4.0. *Procedia CIRP*, 19, 51–56. Available from: https://doi.org/10.1016/j.procir.2014.05.016 [Accessed 09 December 2020].

Solakivi, T., Hofmann, E., Töyli, J., and Ojala, L., 2018. The performance of logistics service providers and the logistics costs of shippers: A comparative study of Finland and Switzerland. *International Journal of Logistics Research and Applications*, 21(4), 444–463.

Stentoft, J., Jensen, K. W., Philipsen, K., and Haug, A., 2019. Drivers and barriers for industry 4.0 readiness and practice: A SME perspective with empirical evidence. Proceedings of the 52nd Hawaii International Conference on System Sciences. Available from: https://doi.org/10.24251/HICSS.2019.619 [Accessed 09 December 2020].

Sung, T. K., 2018. Industry 4.0: A Korea perspective. *Technological Forecasting and Social Change*, 132, 40–45. Available from: https://doi.org/10.1016/j.techfore.2017.11.005 [Accessed 09 December 2020].

Transparency International, 2020. Corruption perceptions index 2019. Available from: www.transparency.org/cpi [Accessed 09 December 2020].

Türkeş, M. C., Oncioiu, I., Aslam, H. D., Marin-Pantelescu, A., Topor, D. I., and Căpuşneanu, S., 2019. Drivers and barriers in using industry 4.0: A perspective of SMEs in Romania. *Processes*, 7(3), 153. Available from: https://doi.org/10.3390/pr7030153 [Accessed 09 December 2020].

UBS, 2016. Extreme automation and connectivity: The global, regional, and investment implications of the Fourth Industrial Revolution. UBS White Paper for the World Economic Forum Annual Meeting.

Urciuoli, L., Hintsa, J., and Ahokas, J., 2013. Drivers and barriers affecting usage of e-Customs-A global survey with customs administrations using multivariate analysis techniques. *Government Information Quarterly*, 30(4), 473–485. Available from: https://doi.org/10.1016/j.giq.2013.06.001 [Accessed 09 December 2020].

Vahti, J., 2020. Sitra mukaan Euroopan digitulevaisuutta rakentavan GAIA-X-projektiin [Sitra joins the GAIA-X project shaping Europe's digital future]. Available from www.sitra.fi/uutiset/sitra-mukaan-euroopan-digitulevaisuutta-rakentavaan-gaia-x-projektiin/ [Accessed 09 December 2020].

Wang, H., Osen, O. L., Li, G., Li, W., Dai, H. N., and Zeng, W., 2015. Big data and industrial internet of things for the maritime industry in northwestern Norway. Proceedings of the TENCON 2015–2015 IEEE Region 10 Conference, 1–4 November 2015 Macao, China, 1–5.

Wellener, P., Dollar, P., Ashton, H., Monck, L., and Hussain, A., 2020. The future of work in manufacturing: What will jobs looks like in the digital era? Available from: https://www2.deloitte.com/us/en/insights/industry/manufacturing/future-of-work-manufacturing-jobs-in-digital-era.html [Accessed 3 December 2020].

WEF, 2016. *The Global Information Technology Report 2016: Innovating in the Digital Economy*. The Global Information Technology Report. Geneva: World Economic Forum.

WEF, 2020. *The Future of Jobs Report 2020*. Geneva: World Economic Forum.

Xu, L. D., Xu, E. L., and Li, L., 2018. Industry 4.0: State of the art and future trends. *International Journal of Production Research*, 56(8), 2941–2962. Available from: https://doi.org/10.1080/00207543.2018.1444806 [Accessed 09 December 2020].

Zhu, Q., and Geng, Y., 2013. Drivers and barriers of extended supply chain practices for energy saving and emission reduction among Chinese manufacturers. *Journal of Cleaner Production*, 40, 6–12. Available from: https://doi.org/10.1016/j.jclepro.2010.09.017 [Accessed 09 December 2020].

4 Small and Medium-Sized Enterprises and the Industry 4.0 Transformation in France

Elena Kornyshova and Judith Barrios

CONTENTS

4.1 INTRODUCTION

The website of the *République numérique* (*Digital Republic*) states that "The Republic of the 21st century will necessarily be digital" (Pour une République numérique, 2016). The Industry 4.0 (I4.0) transformation, as a component of digitalization, changes the functioning of organizations and implies that they should become compatible with this revolution. (Pfohl *et al.*, 2015) defines I4.0 as "the sum of disruptive innovations derived and implemented in a value chain to address the trends of digitalization, automatization, transparency, mobility, modularization, network collaboration and socializing of products and processes." As defined in (Müller *et al.*, 2018), three main dimensions characterize I4.0: the digitalization of processes, smart manufacturing with cyber-physical systems, and "inter-company connectivity between suppliers and customers within the value chain." Different kinds of companies are more or less sensitive to these changes. The study by (Müller *et al.*, 2020) showed that large enterprises can more easily utilize their potential and capacities to develop innovative strategies than small- and medium-sized enterprises (SMEs), which are often in the initial stages of the I4.0 event and just beginning to implement

DOI: 10.1201/9781003165880-4

I4.0 technologies (Müller *et al.*, 2020). SMEs often lack the financial, human, informational, and material resources to succeed in their I4.0 transformations.

It has been a fundamental objective to facilitating the progress of French enterprises in the world of digital transformation. The French government works on programs, policies, and relaunching plans to support enterprise transformation initiatives (Nouvelle France Industrielle, 2017) (Alliance Industrie du Futur, 2015). In this chapter, we investigate the situation in France to answer questions related to how SMEs are supported by the French government and other institutions, that is, what are the drivers, barriers, and opportunities for SMEs, and what are their prospects? We briefly introduce the French economy and describe the situation that French companies, including SMEs, face within the I4.0 transformation. We discuss the digitalization policy in France in order to subsequently analyze the conditions of SMEs in France with regard to progressing through the I4.0 transformation.

This chapter is organized as follows. The second section presents the methodology we used to conduct our study. The third section describes the French background, allowing us to understand the main economic indicators and the situation of SMEs with regard to I4.0. In the fourth section, we detail the digitalization policy in France. The next three sections are dedicated to the I4.0 related drivers, barriers, and opportunities, respectively. The last section concludes the chapter.

4.2 METHODOLOGY

The methodology used to progress through the study of I4.0 development in France consisted of two main steps. The first step included an Internet search of various sources related to I4.0 in France, including official government websites and the websites of the leading business organizations in France. This search was related to three items: French economic indicators; laws, programs, and other initiatives to support I4.0 in France, particularly for SMEs; and finally, topics related to the I4.0 drivers, barriers, and opportunities. We aggregated the obtained information into the corresponding sections. Our second step included semi-structured interviews with three experts to supplement the data found on the internet. The interview was comprised of five questions:

1. Are you familiar with French governmental policies aimed at helping companies and other institutions to follow the I4.0 revolution?
2. What are the I4.0 drivers in France?
3. What are the barriers to the adoption of I4.0 technologies in France?
4. What opportunities are instigated for companies and institutions by I4.0 in France?
5. How do you foresee I4.0's prospects in France?

All the obtained results were synthesized and are presented in the following sections.

4.3 COUNTRY BACKGROUND

In Table 4.1, we present the main economic indicators characterizing France. With a GDP of €2,275,668 million (Country Economy, 2021), France is placed third in

TABLE 4.1
Selected Economic Indicators

Indicator	Unit	Value	Source
Country population	Million	64,993	(International Monetary Fund, 2021a)
GDP 2020	Million €	2,275,668	(Country Economy, 2021)
GDP per capita 2020	€	33,804	(Country Economy, 2021)
Real GDP growth	Annual percent change	6	(International Monetary Fund, 2021a)
2021 projected real GDP	% change	5.5	(International Monetary Fund, 2021a)
Inflation rate, average consumer prices	Annual percent change	0.6	(International Monetary Fund, 2021a)
2021 Projected consumer prices	% change	0.6	(International Monetary Fund, 2021a)

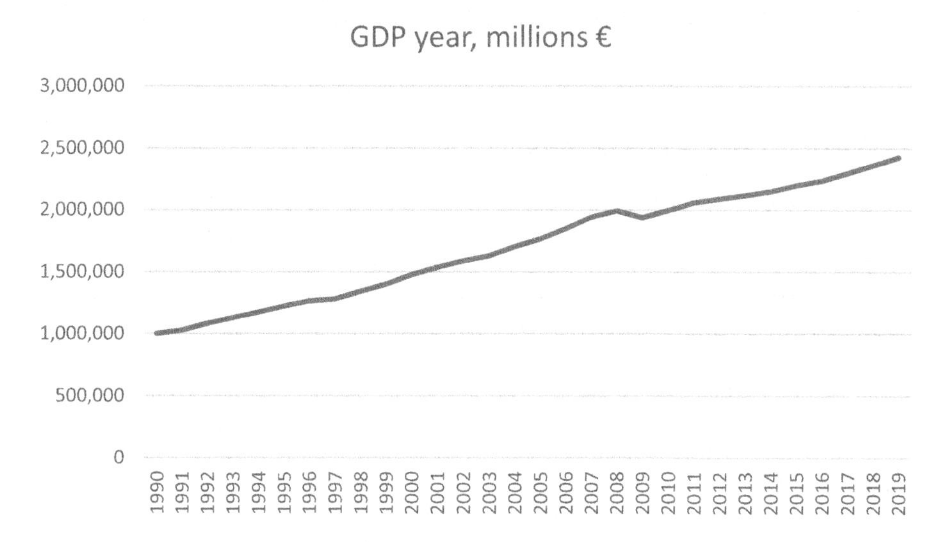

FIGURE 4.1 Gross domestic product progression in France.

Source: Based on data presented in Country Economy (2021)

Europe (after Germany and the United Kingdom) and seventh in the world (World Population Review, 2021). The French population is approximately 65 million with a GDP of €33,804 per capita (Country Economy, 2021). France's economic growth is stable. Figure 4.1 shows the annual progression of the gross domestic product (GDP) for the last three decades. IMF analysts project that the GDP will grow by 5.5 percent in 2021 with a relatively stable inflation rate and consumer prices change (International Monetary Fund, 2021a).

French industry has considerable growth and impact potential. As expressed in World Population Review (2021), "France's economy is a diversified free-market-oriented economy"; its most developed sector is the chemical industry (first place) together with agriculture and tourism. In the world, France occupies sixth place in terms of agricultural production and second place as an exporter of agricultural products (World Population Review, 2021). France ranks first as the most attractive tourist destination in the world (World Population Review, 2021). France takes fifth place in the Fortune Global 500; 28 of the 500 largest companies in the world are French (World Population Review, 2021).

(Le Clainche *et al.*, 2020) presents a summary of the data on French companies in 2018. The number of microenterprises and other SMEs in France is shown in Table 4.2, along with the corresponding number of employees and generated added value. SMEs in France are distributed between sectors in the following manner (cf. Table 4.3). The sectors most present are trade and market services; followed by the industry and building sectors, which are almost at the same level; and, finally, transportation is the least developed sector in France.

In 2019, *GFI Informatique*, recently renamed Inetum (GFI Informatique, 2021) conducted a study with 200 French medium-sized manufacturing enterprises (from 100 employees), which was published in February 2020 (Baromètre Industrie 4.0

TABLE 4.2
Key Data on SMEs in France, 2018

	Microenterprises	SMEs without microenterprises	Total SMEs
Number	3,779,880	148,078	3,927,958
Employees, FTE, thousands	2,421	3,849	6,270
Added Value, billion euros, excluding tax	236	276	512

Source: Not including agricultural and financial companies; based on data from Le Clainche *et al.* (2020)

TABLE 4.3
SME Distribution between Sectors

Sector	Number of companies, thousands	Employees, FTE
Industry	24	772
including manufacturing industry	*23*	*728*
Construction	23	501
Trade	35	799
Transportation	8	243
Market services	45	1,213
Total	**136**	**3,529**

Source: not including microenterprises, nor agricultural, financial, and real estate companies; based on data from Le Clainche *et al.* (2020)

des ETI, 2020). The results of this study show the impressions of I4.0, the appetite for I4.0, an I4.0 risk evaluation, and French companies' maturity with regards to I4.0 (cf. corresponding graphics in Figures 4.2–4.5). The study also demonstrated that the I4.0 transformation has had globally a positive impact (for 39 companies with I4.0 completed or ongoing transformations): for 10 percent, the transformation

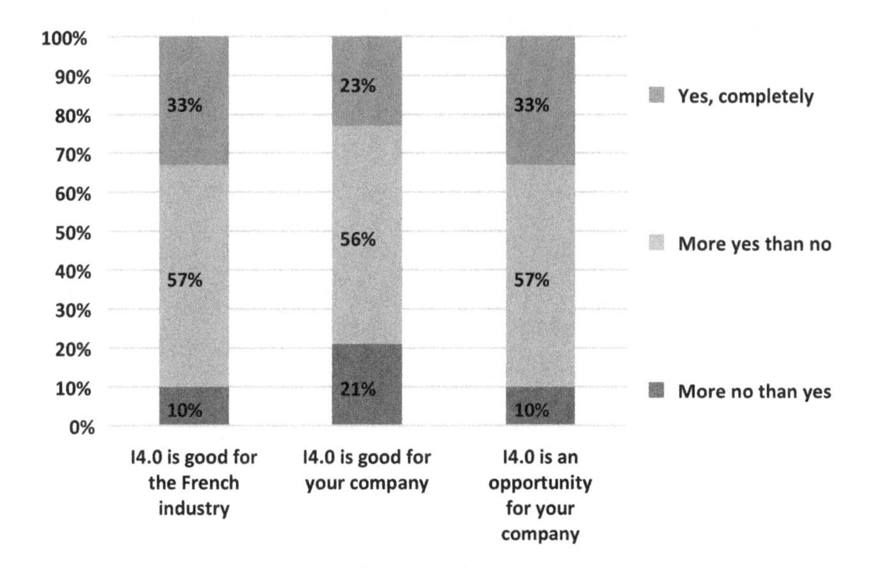

FIGURE 4.2 French companies' impressions of I4.0.

Source: Based on data from Baromètre Industrie 4.0 des ETI (2020)

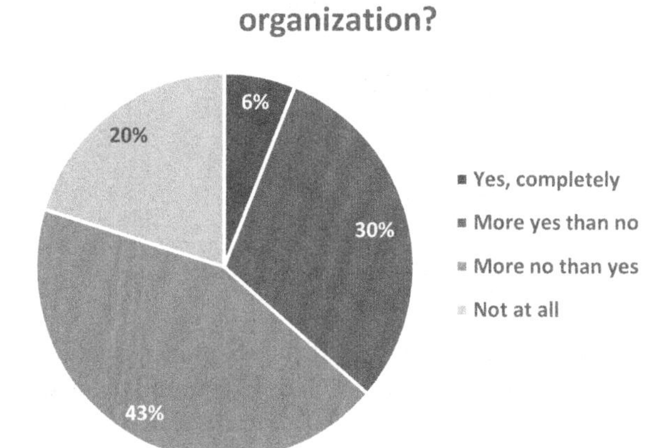

FIGURE 4.3 Perceived risks of I4.0.

Source: Based on data from Baromètre Industrie 4.0 des ETI (2020)

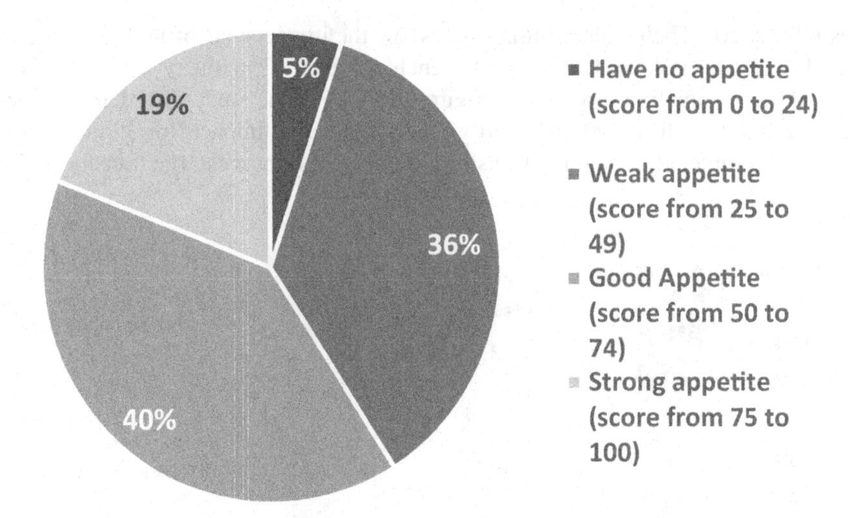

FIGURE 4.4 French manufacturing companies' appetite for I4.0.

Source: Based on data from Baromètre Industrie 4.0 des ETI (2020)

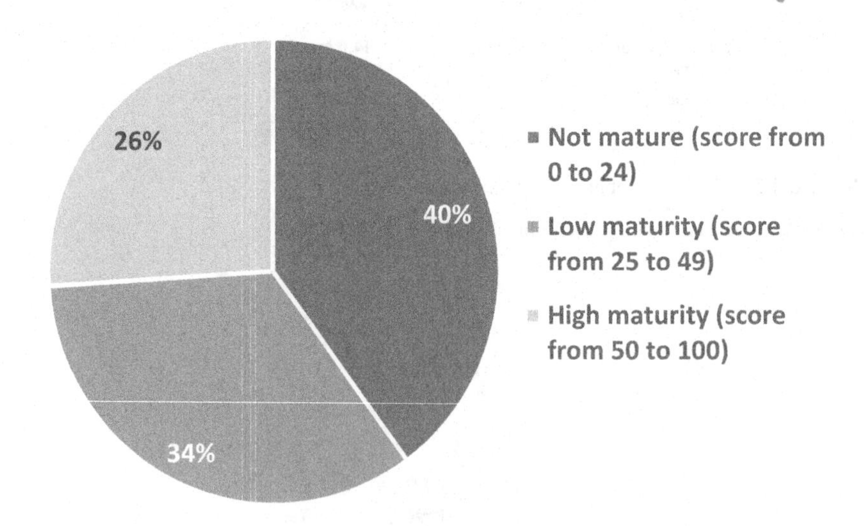

FIGURE 4.5 Maturity of French companies in I4.0.

Source: Based on data from Baromètre Industrie 4.0 des ETI (2020)

has been completely successful; for 76 percent, the transformation has been more of a success than a failure; for 6 percent, it has been a failure; and 8 percent were unable to answer (Baromètre Industrie 4.0 des ETI, 2020). Even though I4.0 is globally valued by these companies, real engagement with I4.0 is not yet anticipated for the majority of them. Within the whole set of the study participants (200 companies), only 37 percent are really engaged in the I4.0 transformation (finished, ongoing, or planned for the next 12 months), the other 63 percent have no intention of commencing any I4.0-related project in the next 12 months (Baromètre Industrie 4.0 des ETI, 2020).

4.4 DIGITALIZATION POLICY IN FRANCE

The French government has been the source of several national initiatives that aim to define enterprise modernization strategies. Generally, these initiatives are implemented by multidisciplinary groups which discuss the directions, priorities, and support programs that ought to be implemented to outline what kind of industry will be needed in the near future.

In 2013, French President F. Hollande launched the *Nouvelle France Industrielle* (New Industrial France) to support French companies in their evolution and to contribute to their better positioning in future markets (Nouvelle France Industrielle, 2017). The goal of this action was to "bring France to the forefront of global competition by writing a new page in its industrial story" (La Nouvelle France Industrielle, 2013). The goal of this action was to stimulate French involvement in the modernization of its industry.

Industrie du futur (Industry of the Future) was launched in April 2015 to assist companies with their modernization and transformation in order to follow the I4.0 revolution (Nouvelle France Industrielle, 2017). This initiative defines the five pillars of the Industry of the Future; these are (1) developing the supply for the industry of the future, (2) supporting companies to move towards the industry of the future, (3) staff training, (4) promoting the industry of the future, and (5) reinforcing European and international cooperation (Nouvelle France Industrielle, 2017). To concretize this programme, *Alliance Industrie du Futur* (Future Industry Alliance) (Alliance Industrie du Futur, 2015) was founded in July 2015 to organize and coordinate different projects, initiatives, and work around the transformation of French industry, including coping with digitalization (Nouvelle France Industrielle, 2017). This alliance includes professional, academic, and technological organizations along with organizations for financing companies. It consists of 32 members overall. The Future Industry Alliance is directly related to the National Council for Industry. The representatives from these organizations participate in six working groups (such as *Vitrines Industrie du futur* – Showcasing the Industry of the Future, *Développement de l'offre technologique du Futur* – Designing the Future Technological Supply, or *Homme et Industrie du Futur* – People and the Industry of the Future) and two transversal actions in order to implement program strategies (Alliance Industrie du Futur, 2015). The Future Industry Alliance is considered to be the key operational player within French industrial policy. It supports French companies in the modernization of their industrial tools and the transformation of their business models through new, digital, and non-digital technologies. As underlined in (European Commission Digital Transformation Monitor, 2017), SMEs and mid-cap companies constitute the target audience of the Alliance.

Law No. 2016–1321 of October 7, 2016 "*Pour une République numérique*" ("For a Digital Republic") aims to promote "the openness and circulation of data and knowledge and to facilitate citizens' access to digital technologies" (Pour une République numérique, 2016). It includes the following provisions: the circulation of data and knowledge, rights protection in the digital society, and access to digital technologies (LOI n° 2016–1321 du 7 octobre 2016 pour une République numérique, 2016).

La French Fab (La French Fab, 2021) was launched in 2017 (Krummenacker, 2020) and initiated by Bruno Le Maire, Minister of the Economy and Finance. The goal of the French Fab is to share best digital practices across French industries and to promote the French industry exports. It is considered to be a new "Made in France."

Another program called PFI4 – *Plateforme France Industrie 4.0* (Platform France I4.0) (Plateforme France Industrie 4.0, 2021) was launched in 2019 to accelerate the deployment of I4.0 within companies. PFI4 has developed I4.0 training modules that are both conceptually and practically robust. It relies on partnerships established with academic reference organizations such as Dauphine Université Paris or Centrale Supélec and on a network of trainers with experience in implementing I4.0. It includes the following:

- Support for companies (financial, material, and so on).
- Training of employees.
- Promotion of the French I4.0 outside of France.
- Launch of different R&D projects related to I4.0.
- Reinforcement of international cooperation especially with regard to different standards.

The digitalization policy in France is also related to several international initiatives. The best known of these is *GAIA-X* (GAIA-X, 2021), which began in 2019 (Bongers *et al.*, 2020). GAIA-X is a European global project to facilitate the creation of European data and artificial intelligence-driven ecosystems to guarantee data sovereignty, develop conceptual foundations for shared data infrastructure, create an ecosystem of users and providers, and establish corresponding structures (Bongers *et al.*, 2020). The French Hub of GAIA-X (Création d'une entité de gouvernance pour Gaia-X, 2021) was opened in 2020 to promote and encourage the GAIA-X activities in France. In addition, French companies benefit from France's reputation for attracting foreign investors. (Business France, 2020a) stated that "In 2019, France became the leading host country for job-creating foreign investments in Europe and remained in first place for industrial and R&D projects."

Finally, it is important to note that our interviewed experts confirmed that the initiatives of the French government are important but at the same time are still somewhat limited:

- *"The French government does care about I4.0. Multiple projects have been launched to support I4.0. However, it is still vague compared, for instance, to Germany which has a more detailed plan and is more advanced in this field. In France, reflections have begun despite being not well defined."*

- *"The government aims at promoting I4.0 to optimize resources, it could be considered a way to invest in order to obtain future economies. This is the case with state modernization programmes such as FTAP (Fond pour la transformation de l'action publique* – Fund for the Transformation of Public Action). *New technologies are considered to be an opportunity to reduce costs and to obtain a better result with fewer expenses."*

4.5 INDUSTRY 4.0 DRIVERS

We have identified the main drivers enabling French companies including SMEs to proceed with the I4.0 transformation based on the following sources: Business France (2020a) details the attractive factors of the French companies; Business France (2020b) synthesizes the key strengths of the French economy and French companies; and European Commission Digital Transformation Monitor (2017) presents the key French drivers related to the Future Industry Alliance.

The I4.0 drivers in France are related to a variety of aspects:

- ***Governmental Programs***. The French government's initiatives, such as *the Future Industry Alliance* or *the French Fab*, help companies to innovate and convert to digitalization (European Commission Digital Transformation Monitor, 2017). In addition, they helped organizations during the COVID-19 crisis with recovery plans for several sectors (Business France, 2020b, Section "France: a Global Leader in Industry"). For instance, the Future Industry Alliance involved different stakeholders (business, research, technology) in activities related to this alliance, hence, in the design and implementation of the Industry 4.0 technologies (European Commission Digital Transformation Monitor, 2017). National initiatives are also supported by French regions that increase the global resources available to help French companies.
- ***Research and Innovation***. In 2018, France was ranked sixth in the world in regard to R&D expenditure (Business France, 2020a). In 2020, France was ranked third in the world and first in Europe in the list of the Top 100 Global Innovators produced by *Clarivate* with five French organizations included in this list (Business France, 2020b, Section "France: An Innovative Powerhouse"). Our experts discussed scientific and engineering excellence in France and mentioned the "*network of excellent tertiary engineering schools such as Plateforme Saclay*" (Plug in Labs, 2021). They also stated that "*Several competencies such as plane welders represent a very particular expertise and are popular even outside of France.*"
- ***Quality of Infrastructures***. High-quality transport and communication infrastructures, and an extensive broadband network in France are key factors required for the development of the competitiveness of French companies (Business France, 2020a). One of the experts concurred with this idea: "*Very good transport infrastructures (roads, railways, air connexions,*

etc.) constitute a kind of physical facilitator for I4.0"; "*The same for the phone connections. The development of networks including 5G contributes to I4.0.*"

- ***Taste for Entrepreneurship and Motivation for Innovation***. In 2019, France was classified fourteenth in the Global Entrepreneurship Index and fourth for Entrepreneurial Aspirations (Ács *et al.*, 2019). Entrepreneurial aspiration is seen as "the early-stage entrepreneur's effort to introduce new products and/or services, develop new production processes" (Ács *et al.*, 2019). This driver was also cited by our experts: "*Sometimes this is about the desire for new technologies,*" "*the French spirit of endeavours/achievement,*" "*a will to make extraordinary things like the Concorde plane, the Airbus, TGV, a will to innovate even if sometimes this will do not proceed, the ideas are under-exploited afterward compared to America, for instance.*"

4.6 INDUSTRY 4.0 BARRIERS

In France, the barriers to the development of I4.0 are numerous. For instance, the report by the National Productivity Board of France (National Productivity Board, 2019) provides a good overview of the situation. This report confirms the high level of productivity in France; however, it indicates that productivity has decreased since the late 1990s. This decrease is related to several factors that could also hinder the I4.0 transformation, including support of SMEs.

- ***Regulatory Barriers***. Several market regulations in France represent restrictions or constraints for French companies and hinder investment in information and communication technology: the "existence of monopoly rents in several sectors, . . ., reporting requirements, registration with a corporation or sector-level chamber, the cost of acquiring a licence, the complexity and ambiguity of regulations and procedures" (National Productivity Board, 2019). In addition, France has a high tax level in general (Business France, 2020a), albeit the French government has undertaken a number of actions to facilitate the functioning of companies, especially SMEs (these measures are detailed in the next section). Our experts expressed the same ideas: "*Taxes on production are very high compared to other European countries*"; "*In France there are multiple taxes and especially taxes on production and storage*"; "*There are laws to stimulate I4.0 but laws are also barriers as now many industrial activities are outsourced outside of France because of economic issues induced partially by the existing laws.*"
- ***Relatively Low Level of French Workforce Skills***. (National Productivity Board, 2019) stipulates that "the skills of the French workforce are below the OECD average and that there is hardly any sign of improvement." In addition, there is a lack of correlation between the skills being attained and the skills required by the labor market (European Commission Digital Transformation Monitor, 2017). One of the interviewed experts mentioned the problem of staff availability, that is, "*a shortage of employees qualified*

for I4.0." "The current level of unemployment in France is about 9%. The education does not fit the employment market requirements including professions related to I4.0. In addition, with the current COVID-19 situation, people want to go back to work but often they cannot due to the lack or incompatibility of their education." "The need for I4.0-conversant employees will only grow."

- ***Social Disparities***. "The French education system is characterized by a greater skills gap between people from different social backgrounds relative to other countries" (National Productivity Board, 2019). Lifelong learning options are limited for the most vulnerable employees. Our experts confirmed this idea: *"Industry 4.0 pushes people up. However, French society is very stratified and rigid. This applies to private companies and public administrations. COVID-19 showed that even hospitals have minimal communication between their different services. Disparities are also between hierarchical levels and even between social groups. Barriers are very strict and there is almost a social distortion. I4.0 requires that all participants work together."*

Other barriers suggested by the interviewed experts consist of the following:

- ***Trade Union Actions***. Trade unions sometimes represent a barrier as they are often opposed to changes in working methods: *"I4.0 could be seen as a substitution of people by machines"; "The very high protection of employment does not contribute to replacing human workstations with machines often required by the I4.0 transformation."*
- ***Outsourced Activities***. *"Many industries are outsourced from France and have even disappeared. This decreases the number of potential cases for I4.0 development."*
- ***Weak Diffusion of Digital Technologies.*** *"Insufficient dissemination of digital technologies through industry"; "R&D activities are not present in all companies."*

4.7 INDUSTRY 4.0 OPPORTUNITIES

The main I4.0 opportunities for French companies are related to different government projects and tax reductions. The French government supports companies and individuals by "enforcing contracts, starting a business, trading across borders, etc." (Business France, 2020a). France scores well in these categories compared to other countries and "holds a middle-ranking position among the major economies for business environments" (Business France, 2020a).

- ***Government Projects.*** The website of the *Plan de Relance* (Relaunch Plan) (Plan de Relance, 2021) details different projects for SMEs, including very small enterprises. Some examples are presented in the following:
  - *Aide à l'investissement de transformation vers l'industrie du futur* (Support for investment in the transformation to the industry of the

Future) (Aide à l'investissement de transformation vers l'industrie du futur, 2020). This aid covers 40 percent of the total investment in new technologies (robotics, virtual or augmented reality, etc.) with a limit of €200,000 or €800,000 depending on certain factors.
- *Aide pour la maîtrise et la diffusion numérique dans le cadre de "IA Booster*" (Support for digital literacy and dissemination in the framework of the "Artificial Intelligence Booster") (Aide pour la maîtrise et la diffusion numérique dans le cadre de "IA Booster," 2020). This project offers help with integrating artificial intelligence solutions to reduce manufacturing costs, enhance competitiveness, and upgrade systems. Concrete measures will be available in the near future.
- *Aides France Num pour la transformation numérique* (France Num aid for digital transformation) (Aides France Num pour la transformation numérique, 2020). The project offers a cheque of €500 to very small and SMEs in the commercial and handicrafts sectors along with other assistance (recommendations, diagnosis, etc.).

- ***Tax Reductions.*** France "offers attractive tax rates in various fields: R&D, digital services, SMEs and innovative new companies, intellectual property income, special expatriate tax system, etc." (Business France, 2020b, Section "France: Separating Fact from Fiction"). Among others, several measures are related to manufacturing, R&D, are special for SMEs and for the current COVID-19 crisis.
    - *Manufacturing taxes and dues reductions* (Baisse des impôts de production, 2021): the reduction by half of the *CVAE – Cotisation sur la Valeur Ajoutée des Entreprises* (Corporate Value Added Tax), the reduction by half of the *CFE – Cotisation Foncière des Entreprises* (Real Estate Contribution) and of the *TFPB – Taxe Foncière sur les Propriétés Bâties* (Tax on Built Properties) for industrial establishments, the reduction from 3 percent to 2 percent of the capping rate of the *CET – Cotisation Economique Territoriale* (Territorial Economic Contribution).
    - *R&D activities*: *CIR – Crédit d'impôt recherche* (Research tax credit) (Crédit d'impôt recherche, 2021) has been offered since 1983 to French companies to motivate them to invest in research and *CIFRE – Convention industrielle de formation par la recherch*e (Industrial convention for training through research) (Le dispositif CIFRE, 2021) allows students to prepare a Ph.D. project in relation to a company.
    - *Tax Reductions for SMEs.* SMEs receive special treatment in France: "reduced tax rates of 15% and 28% depending on profits (from €0 to €38,120 and from €38,120 to €500,000)" (Business France, 2020a).
    - *COVID-19 Crisis Measures.* The International Monetary Fund also notes that the French government has responded to the COVID-19 crisis with several fiscal benefits: "France's fiscal response to the crisis was timely, flexible, and appropriately calibrated to the size of the shock . . . the composition of the fiscal package was flexibly adjusted,

> with additional spending and foregone revenue as the crisis unfolded" (International Monetary Fund, 2021b).

To progress to I4.0 transformation, French companies are also being assisted by different organizations. The aforementioned study (Baromètre Industrie 4.0 des ETI, 2020) showed that companies have support from different kinds of partners. Of 76 participating companies (with finished, ongoing, or planned I4.0 transformation), 51 percent were helped by IT consulting companies, 38 percent by public bodies (state and local authorities, research centers), and 37 percent by start-ups (a combination of different partners is possible).

French SMEs and other companies benefit from collaborations of various natures:

- ***International Collaborations***. French SMEs can take advantage of opportunities related to international collaborations. For instance, the *GAIA-X project* supports SMEs in the following areas (GAIA-X, 2021): collaborative condition monitoring, smart manufacturing, supply chain collaboration in a connected industry, shared production, predictive maintenance, and so on.
- ***French Associations***. French associations, business groups, and professionals analyze, guide, support, and follow organizational transformation initiatives. The best known is *CIGREF – Club informatique des grandes entreprises françaises* (Informatics Club of Large French Businesses) (CIGREF, 2020). This association represents many French companies as well as some public administrations that utilize digital solutions and services. The association has brought together 150 members who combine their thinking and experience/backgrounds in digital issues. Its mission is to develop the capacity of large companies to integrate and master digital technology. *CAP Digital* (Cap Digital, 2020) is another example of an association that assembles many players in the field of innovation such as start-ups, research laboratories; small, medium, and large companies; schools; universities; territories; and investors. The objective of these associations is to facilitate the collaboration between digital economic sectors, specialists in technology, and sustainability in order to produce innovative solutions.

Other opportunities detected by the experts are listed here:

- ***Slowdown of the productivity increase.*** "*With productivity slowing down for several years, digital industrialization is an opportunity to improve productivity.*"
- ***Staff availability***. "*Because of unemployment, there are many persons who are available to integrate into the I4.0 revolution. For I2.0 and I3.0, each time the revolution was made possible/even facilitated by the migration of people from the countryside to the cities. It is the same for I4.0, available workers are essential to progress. Even if I4.0 somehow contributes to the removal of people, they will find another activity. For instance, in*

a pharmacy a new robotic chain will allow the quicker distribution of the ordered drugs, hence, people will be freed from distribution to provide better service at the reception point."

4.8 CONCLUSION

Currently, several prospects are open to French companies, including SMEs, to enhance their competitiveness in moving towards the I4.0 transformation. The study of the situation in France has shown that several drivers (such as reducing costs) and barriers (lack of skills) are shared with other countries (Horváth and Szabó, 2019; Moeuf *et al.*, 2020), but at the same time, French companies face several additional difficulties related to regulatory barriers, social disparities, trade union actions, and so on. A summary of the main I4.0 drivers, barriers, and opportunities reported in this chapter is presented in Table 4.4.

The experts have stated that "*The passage to I4.0 is mandatory.*" Companies, especially SMEs, should be helped in this transformation, and multiple actions are ongoing in this regard. Sometimes the I4.0 evolution is guided by "*a frenzy for technology, not by a real need,*" and the target vision of I4.0 is not really defined: "*We should continue, but where? How?*" Sometimes there is a difficulty focusing on usages because of "*the vision of 'technology first' which hides users' needs.*" "*Now there are many obstacles,*" however, "*France has a good web of digital knowledge that will help.*" Accordingly, our interviewed experts recognize the difficulties that

TABLE 4.4
Summary of French Drivers, Barriers, and Opportunities

Drivers	Barriers	Opportunities
• Governmental programmes: – Nouvelle France Industrielle (New Industrial France), – Industrie du futur (Industry of the Future) and Alliance Industrie du Futur (Future Industry Alliance), – Law "Pour une République numérique" ("For a Digital Republic"), – La French Fab, – Plateforme France Industrie 4.0 (Platform France I4.0) • Research and innovation, scientific and engineering excellence • Quality of infrastructures • Taste for entrepreneurship and motivation for innovation	• Regulatory barriers • Relatively low level of French workforce skills • Social disparities • Trade union actions • Outsourced activities • Weak diffusion of digital technologies	• Government projects, such as Plan de relance (Relaunch Plan) • Tax reductions: – Manufacturing taxes and dues reductions (CVAE, CFE, TFPB, CET) – Tax facilities related to R&D activities, such as CIR and CIFRE – Tax reductions for SMEs – COVID-19 crisis measures • International collaborations, such as GAIA-X • French associations, such as CIGREF or CAP Digital • Slowing of the productivity increase • Staff availability

French companies have, but they also believe in a positive outlook for I4.0 in France: "*There are several difficulties, but they should be overcome*"; "*The prospects are good and should be good*"; "*Several companies will adopt these technologies very quickly, and we hope that others will follow*"; "*A to-do is to enhance the general culture of people*"; "*Opportunities are huge; multiple companies have succeeded in the I4.0 revolution despite obstacles, thus, prospects are very good.*"

Many French initiatives supported by European and global projects are helping French SMEs and other companies to advance in the direction of digitalization. The main drivers represent government initiatives seeking to lower the barriers that obstruct the digital transformation process; at the same time, these initiatives unleash a range of encouraging possibilities for French companies. Thus, we believe that different drivers and opportunities will progressively allow the removal of the existing barriers to motivate French enterprises to progress to an I4.0 transformation process and contribute to their success.

4.9 ACKNOWLEDGEMENTS

The authors would like to thank the editors for the idea to produce this book and for their invitation to participate in its writing. The authors would also like to express their thanks to the following experts: Samia Bouzefrane, Professor at the *Conservatoire National des Arts et Métiers* (Cnam) of Paris and Lead Expert at the Ministry of Higher Education, Research and Innovation; Lionel Icard, Engineer and Expert in Banking Information Systems; and Jean-Marie Rousseau, Head of Department at the *Institut de Radioprotection et de Sûreté Nucléaire* (IRSN) for sharing their points of view and for contributing to this chapter.

REFERENCES

Ács, Z.J., Szerb, L., Lafuente, E., and Márkus, G. (2019). *Global Entrepreneurship Index, 2019*. Available from: https://thegedi.org/global-entrepreneurship-and-development-index/ [Accessed March 2021].

Aide à l'investissement de transformation vers l'industrie du futur. Available from: www.economie.gouv.fr/plan-de-relance/profils/entreprises/aide-investissement-industrie-du-futur [Accessed November 2020].

Aide pour la maîtrise et la diffusion numérique dans le cadre de "IA Booster." Available from: www.economie.gouv.fr/plan-de-relance/profils/entreprises/aide-maitrise-diffusion-numerique-iabooster [Accessed November 2020].

Aides France Num pour la transformation numérique. Available from: www.economie.gouv.fr/plan-de-relance/profils/entreprises/aides-francenum-transformation-numerique [Accessed November 2020].

Alliance Industrie du Futur (2015). Available from: www.industrie-dufutur.org/aif/ [Accessed November 2020].

Baisse des impôts de production. Available from: www.economie.gouv.fr/plan-de-relance/profils/entreprises/baisse-impots-production [Accessed February 2021].

Baromètre Industrie 4.0 des ETI. February (2020). Available from: https://gfi.world/fr-fr/presse/publication/36-barometre-industrie-4-0-des-eti [Accessed January 2021].

Bongers, A., Chidambaram, R., Feld, T., Garloff, K., Ingenrieth LLM, F., Jochem, M., Maier, B., Marsch, C., Marti, A.P., Otto, B., Ottradovetz, K., Parshin, V., Pfrommer, J., Plass, C., Reinhardt,

R., Grossón, M.S., Schmieg, A., Schoppenhauer, R., Stark, J., Steinbuss, S., Strnadl, C.F., Tesone, R., Weiss, A., Weiss, C., Wessel, S., and Wilfling, S. May (2020). *GAIA-X: Driver of Digital Innovation in Europe. Featuring the Next Generation of Data Infrastructure.* Available from: www.data-infrastructure.eu/GAIAX/Redaktion/EN/Publications/gaia-x-driver-of-digital-innovation-in-europe.html [Accessed March 2021].

Business France. December (2020a). *France Attractiveness Scoreboard 2020.* Available from: https://investinfrance.fr/wp-content/uploads/2017/08/TBA_2020_Rapport_UK.pdf [Accessed March 2021].

Business France. September (2020b). *2020 Kits: Showcasing the French Economy.* Available from: www.businessfrance.fr/en/discover-France-media-Kits-showcasing-the-French-economy [Accessed March 2021].

Cap Digital. Available from: www.capdigital.com/en/ [Accessed November 2020].

CIGREF. Available from: www.cigref.fr/ [Accessed November 2020].

Country Economy: France – PIB – Produit Intérieur Brut (2021). Available from: https://fr.countryeconomy.com/gouvernement/pib/france [Accessed February 2021].

Création d'une entité de gouvernance pour Gaia-X. Available from: www.entreprises.gouv.fr/fr/actualites/creation-d-entite-de-gouvernance-pour-gaia-x [Accessed February 2021].

Crédit d'impôt recherche. Available from: www.service-public.fr/professionnels-entreprises/vosdroits/F23533 [Accessed February 2021].

European Commission Digital Transformation Monitor. January (2017). *France: Industrie du Futur.* Available from: https://ec.europa.eu/growth/tools-databases/dem/monitor/sites/default/files/DTM_Industrie%20du%20Futur%20v1.pdf [Accessed March 2021].

GAIA-X. (2021). GAIA-X: A Federated Data Infrastructure for Europe. Available from: www.data-infrastructure.eu/GAIAX/Navigation/EN/Home/home.html [Accessed February 2021].

GFI Informatique. Available from: https://gfi.world/fr-fr/ [Accessed March 2021].

Horváth, D., and Szabó, R.Z. (2019). Driving Forces and Barriers of Industry 4.0: Do Multinational and Small and Medium-Sized Companies Have Equal Opportunities? *Technological Forecasting and Social Change*, 146, pp. 119–132.

International Monetary Fund, World Economic Outlook Update. January (2021a). Available from: www.imf.org/en/Publications/WEO/Issues/2021/01/26/2021-world-economic-outlook-update [Accessed February 2021].

International Monetary Fund, France: 2020 Article IV Consultation-Press Release; Staff Report; and Statement by the Executive Director for France. January (2021b). Available from: www.imf.org/en/Publications/CR/Issues/2021/01/15/France-2020-Article-IV-Consultation-Press-Release-Staff-Report-and-Statement-by-the-50022 [Accessed February 2021].

Krummenacker, M. April (2020). *France's Paying Strategy in the Industry of the Future.* Industry & Cleantech, Invest in France. Avaialable from: https://world.businessfrance.fr/nordic/2020/04/23/frances-paying-strategy-in-the-industry-of-the-future/ [Accessed March 2021].

La French Fab. Available from: www.gouvernement.fr/action/pour-une-republique-numerique [Accessed March 2021].

La Nouvelle France Industrielle (2013). Available from: www.gouvernement.fr/action/la-nouvelle-france-industrielle [Accessed November 2020].

Le Clainche, L., Morénillas, N., and Sklénard, G. December (2020). *Les entreprises en France.* Édition 2020. Available from: www.insee.fr/fr/statistiques/4987235 [Accessed February 2021].

Le dispositif CIFRE. (2021). Available from: www.anrt.asso.fr/fr/le-dispositif-cifre-7844 [Accessed February 2021].

LOI n° 2016–1321 du 7 octobre 2016 pour une République numérique (1). Octobre (2016). Available from: www.legifrance.gouv.fr/jorf/id/JORFTEXT000033202746?r=jEB4fa5fpn [Accessed November 2020].

Moeuf, A., Lamouri, S., Pellerin, R., Tamayo-Giraldo, S., Tobon-Valencia, E., and Eburdy, R. (2020). Identification of Critical Success Factors, Risks and Opportunities of Industry 4.0 in SMEs. *International Journal of Production Research*, 58(5), pp. 1384–1400.

Müller, J.M., Buliga, O., and Voigt, K.I. (2018). Fortune Favors the Prepared: How SMEs Approach Business Model Innovations in Industry 4.0. *Technological Forecasting and Social Change*, 132, pp. 2–17.

Müller, J.M., Buliga, O., and Voigt, K.I. (2020). The Role of Absorptive Capacity and Innovation Strategy in the Design of Industry 4.0 Business Models-A Comparison Between SMEs and Large Enterprises. *European Management Journal*. doi:10.1016/j.emj.2020.01.002.

National Productivity Board. July (2019). *Productivity and Competitiveness: Where Does France Stand in the Euro Zone?* Report of National Productivity Board. Available from: www.strategie.gouv.fr/sites/strategie.gouv.fr/files/atoms/files/en_1errapportcnp-10july-final.pdf [Accessed November 2020].

Nouvelle France Industrielle: *Construire l'industrie française du futur*. Janvier (2017). Available from: www.entreprises.gouv.fr/files/files/directions_services/politique-et-enjeux/nouvelle-france-industrielle/NFI-DP-janvier-2017-version-FR.pdf [Accessed November 2020].

Pfohl, H.-C., Yahsi, B., and Kurnaz, T. (2015). The Impact of Industry 4.0 on the Supply Chain. In: *Proceedings of the Hamburg International Conference of Logistics (HICL)*, Vol. 20, epubli GmbH, Berlin, pp. 31–58.

Plan de relance (2021). Available from: www.economie.gouv.fr/plan-de-relance/profils/entreprises [Accessed February 2021].

Plateforme France Industrie 4.0. Available from: https://pfi4.com/ [Accessed February 2021].

Plug in Labs. Available from: www.pluginlabs-universiteparissaclay.fr/en [Accessed Februry 2021].

Pour une République numérique (2016). Available from: www.gouvernement.fr/action/pour-une-republique-numerique [Accessed November 2020].

World Population Review (2021). *GDP Ranked by Country 2021*. Available from: https://worldpopulationreview.com/countries/countries-by-gdp [Accessed February 2021].

5 The Case of Industry 4.0 with Hungarian SMEs

Róbert Marciniak, Péter Móricz, and Krisztina Demeter

CONTENTS

5.1 INTRODUCTION

The rapid pace of Industry 4.0 or, in more general terms, digital innovation is one of the key challenges small and medium enterprises (SMEs) face today. Industry 4.0 means a technological framework for digital technologies that could be successfully utilized as Industry 4.0 spans the entire enterprise value chain and may in some cases go beyond by integrating the corporate value chain with the supply chain (Szerb, Komlósi and Páger, 2020). There are many research studies about the motivations, drivers, and barriers of Hungarian enterprises regarding the adoption of Industry 4.0 (Juhász, 2018; Nagy et al., 2018; Szalavetz, 2018; Horváth and Szabó, 2019; Megyeri and Somosi, 2019; Halmosi, 2020). The new digital technologies may have the most important impact on the employment structure, but Industry 4.0 may also affect how enterprises could integrate into global production chains (Éltető, 2019).

The COVID-19 pandemic has accelerated development processes, shifting the focus of efforts towards digital solutions. However, digital transformation as a technology-based investment process is capital-intense that may be a significant challenge for SMEs (Götz et al., 2020). The Hungarian government has long operated an ecosystem that supports the development of SMEs (Borbás, 2014). But the digital transformation process created a new situation that requires greater adaptation

DOI: 10.1201/9781003165880-5

from SMEs and provides the opportunity for creating pure digital enterprises (Szerb, Komlósi and Páger, 2020).

In order to help SMEs in the successful adoption of new digital technologies, it is critical to evaluate and develop the economic ecosystem (Essakly, Wichmann and Spengler, 2019; Abonyi, Czvetkó and Honti, 2020; Szalavetz, 2020). The Hungarian SME sector has long shown lagging in terms of innovation and skills development (Holicza and Tokody, 2016). In the last years, Horváth and Szabó (2019) analyzed some corporate Industry 4.0 projects in Hungary. They identified the shortage of the digitally skilled and educated workforce as the most significant challenge to the fulfilment of Industry 4.0 in Hungary. They state that many elements of the Hungarian entrepreneurial ecosystem require further development in order to create and operate a much more favorable environment for companies using new technology. According to Szerb, Komlósi and Páger (2020), sometimes the current ecosystem is hindering, rather than supporting the emergence of new technologies. They identified six elements as relative weaknesses: (1) human capital and education, (2) formal institutions, regulation and taxation, (3) culture and informal institutions, (4) knowledge production and dissemination (digital), (5) financing, and (6) networking and support (Szerb, Komlósi and Páger, 2020).

As for expectations, based on a large-scale international survey (Szász et al., 2020) supplemented with a literature review, SMEs usually make less effort to invest in Industry 4.0 projects than large ones. Furthermore, less competitive countries, especially countries from Central and Eastern Europe lag behind in these efforts. Nevertheless, these investments improve cost, quality, delivery, and flexibility performance.

In this chapter, we provide an overview of the status and progress of Hungarian SMEs towards digital transformation relying on selected data from a competitiveness survey carried out in Hungary. First, the country background is described shortly including the status of SMEs and the competitiveness of the country. Then, governmental actions for digitalization are introduced. Next, the methodology and data characteristics are discussed. Finally, the results of the analysis and conclusions follow.

5.2 COUNTRY BACKGROUND

According to the Hungarian National Bank's 2020 Competitiveness Report, the real productivity of Hungarian SMEs has increased by about 30 percent since 2010 (Hergár, 2020). Nevertheless, they are still lagging behind large companies in terms of both productivity and wages (Hope, 2019).

In the period 2014–2016, the share of innovative SMEs was 27.9 percent, the EU average was 49.5 percent (Hope, 2019). However, large companies in Hungary are not innovative, either. The value-added of the Hungarian SME sector in 2018 was the highest among the Visegrád countries (Czech Republic, Hungary, Poland, and Slovakia) in the knowledge-intensive sectors (22.2 percent), but altogether, in 2020, the innovation capacity of the Hungarian SME sector is in the last quarter of the EU member states, significantly below the EU average (Hergár, 2020).

Among the Visegrád countries, the Hungarian SME sector is in last place regarding the use of advanced technologies such as big data, 3D printing, or industrial robots (Hergár, 2020). Besides, IT security and organizational innovation are areas in which they lag behind.

According to current research conducted by eNET and Telekom in June of 2020, only a third of SMEs were operating smoothly during the restrictions caused by COVID-19. Half of the SMEs have managed through the pandemic with either reduced capacity or reduced scope of activities, a third said they have not affected by the situation, but 11 percent were forced to suspend their operations. One percent of them were terminated. At the same time, companies that used more digital solutions were the least affected by the turmoil of the crisis. One-third of businesses started using applications that support online teamwork during this period. Also, one-third of businesses have changed the way they do business, and nearly one third have replaced office work by the home office. However, only 21 percent of the companies surveyed have introduced at least one of the digital solutions such as social media presence, website, webshop, online invoicing, or cashless payment (Telekom, 2020).

Some characteristics of SMEs in Hungary compared to EU28 averages are summarized in Table 5.1. Based on the table, Hungarian SMEs are similar to SMEs elsewhere in Europe. The biggest differences are detectable in the ratio of high-profile companies and productivity, both areas where Hungarian SMEs are lagging.

To measure digital competitiveness, several international indicators examine a country's digital performance based on different perspectives. Hungary's position within the EU28 countries is shown in Table 5.2. Overall, the digital environment for SMEs is not supportive. In all but two indicators Hungary is among the last five countries.

TABLE 5.1
Indicators of Hungarian SMEs Compared with EU28 Averages

Indicators	Hungarian SMEs	Average in EU28 SMEs
Share of SMEs of all enterprises	99.8%	99.8%
Share of microenterprises within SMEs	94.1%	93%
SMEs' employment share of total employment	68.3%	66.6%
GDP production of SMEs	52.5%	54.5%
Share of Exporting SMEs	6.7%	Not available
Share of SMEs with knowledge-intensive service and high-tech manufacturing profile	28.4%	33.3%
Total value-added of SMEs	54.1%	56.4%
Average number of people employed by SMEs	3.3	3.9
Productivity of SMEs (Euro/Employee)	19 800	44 600

Source: EC, 2019; EC-ESIF, 2020; Hergár, 2020

TABLE 5.2
Hungary's Positions in Various International Digital Performance Indices

Indices for countries' digital performance	Hungary's relative position among 28 EU members
eGDI (eGovernment Development Index) in 2020	27th position
eGov (eGovernment Benchmark) in 2020	18th position
DESI (Digital Economy and Society Index) in 2019	21st position
GCI (Global Competitiveness Index) in 2019	24th position
WDCR (World Digital Competitiveness Ranking) in 2020	24th position
EIDES (The European Index of Digital Entrepreneurship Systems) in 2019	24th position
IDI (ITU ICT Development Index) in 2017	24th position

Source: Autio et al., 2019; European Commission, 2019; ITU, 2017; IMD, 2020; Schwab, 2019; van der Linden et al., 2020; Zhu et al., 2020

5.3 DIGITALIZATION POLICIES OF HUNGARY

During the 2014–2020 EU budget term, Hungary invested 9.7 percent of the total allocation of EU Funds in enhancing the competitiveness of SMEs (EC-ESIF, 2020). To combine digitalization and competitiveness efforts, the Hungarian government created several programs to facilitate digitalization, as shown in Figure 5.1.

According to the Association of IT Companies (IVSZ) and Microsoft research study in 2019 in Hungary, the ICT sector accounts for at least 20 percent of the total Gross Value Added, and the share of people employed in the digital economy already reaches 17 percent and could reach 25 percent of the Hungarian GDP (IVSZ and Microsoft, 2019). In order to enhance its share, the Hungarian government formulated its digitalization plans and measures in long-term digitalization strategies. In 2014, the Hungarian government adopted the National Info-communication Strategy (NIS), which defined the most important projects in the sector until 2020, including the development of digital competencies, broadband internet, and the transformation of IT training. NIS has been replaced by the National Digitalization Strategy (NDS), adopted in autumn 2020, starting in 2021 (Hungarian Insider, 2020a). The strategy is based on four pillars – digital infrastructure, digital competence, digital economy, and digital state. The overarching goal of the strategy is for Hungary to make concerted efforts to promote digitalization in the fields of economy, education, R&D innovation, and public administration, which will make a significant contribution to improving the country's competitiveness and people's well-being. The most important goal of the new strategy is to make Hungary one of the ten leading EU countries in digitalization by 2030 (Hungarian Insider, 2020a).

Besides NIS and NDS, there are other sectoral or functional strategies, which are linked to the digitalization and are brought together by the Digital Success Program (DJP) (https://digitalisjoletprogram.hu/en). It was launched in 2015 as DJP 1.0 and then continued with DJP 2.0 two years later. DJP programs have carried out more

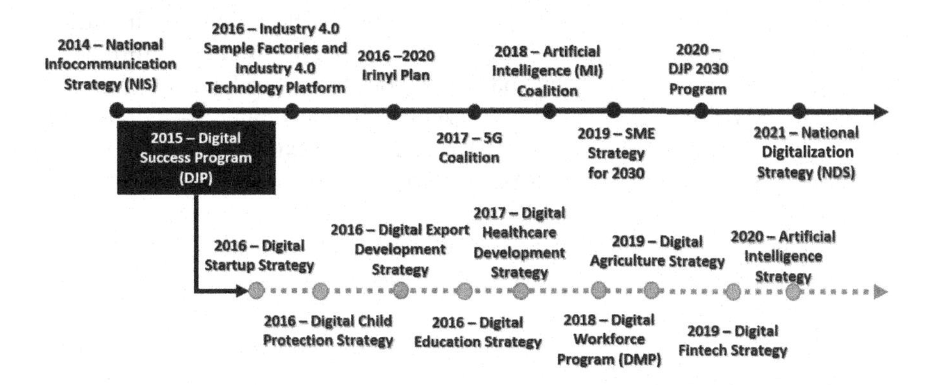

FIGURE 5.1 Different digitalization strategies and frameworks 2014–2020, Hungary.

than 100 projects in dozens of areas. In this framework, various sectoral digital strategies (agriculture, education, export development, child protection, public collection, start-up, fintech, health industry development) have also been developed. The Digital Workforce Program (DMP), as part of DJP, aims to address the shortage of digitally skilled workers, which was named one of the most important obstacles to the development of the digital economy. The goal of the program is to train at least another 20,000 IT professionals within three years besides those trained by the existing educational system. The DJP 2030, adopted in 2020, consolidates the tasks of the next ten years into a single framework, considering the government's tasks related to the development of the country's digital ecosystem into a holistic and structured system.

In parallel with strategies of DJP, other initiatives and programs have also been adopted in Hungary, like Industry 4.0 Sample Factories and an Industry 4.0 Technology Platform. They support the development of domestic manufacturing SMEs and the technological, industrial automation and control solutions of Industry 4.0 production systems. The Irinyi plan (Hungarian Insider, 2020b), in operation since 2016, provides grants for SMEs for developing and launching new products on the world market. 5G Coalition aimed for Hungary to be one of the first countries in the world that introduce and implement the 5G technology. The Artificial Intelligence (MI) Coalition aims to define the directions and framework for the development of artificial intelligence in Hungary, providing a permanent professional and cooperation forum for business, research, academia, and government (https://ai-hungary.com/en).

At the end of 2019, the Hungarian government has drawn up an SME strategy for 2030 to strengthen Hungarian SMEs (Istrate, 2019). The strategy is based on seven pillars, in which one is strengthening the innovation and digital performance of SMEs by spreading the application of technological innovations, developing the digital competences of SMEs. This development policy must ensure that, in adapting to digitalization, SMEs should think besides the digital switchover about changing their place in the international value chains and cooperation. One of the flagship measures of the strategy is the National Entrepreneurship Mentoring Program, the aim of which is to strengthen entrepreneurial awareness and activity in Hungary.

In the last decade, beyond the programs cited earlier, Hungary has introduced several actions to improve SMEs' skills and innovation capabilities. The National Research, Development and Innovation Office had tender calls specifically for SME digitalization; the government provides interest-free loans for technology investments and innovation; the Hungarian Export Promotion Agency (HEPA) also support innovation in SMEs to build up their innovation capacities. (EC, 2019).

Seeing the ambitious strategies, programs, and actions, it might be interesting to get to know how managers understand their barriers, drivers, and opportunities in their digitalization path.

5.4 METHODOLOGY

Our analysis is based on a survey designed by the Competitiveness Research Centre at Corvinus University of Budapest (Chikán, 2008). The applied questionnaire is divided into five parts: (1) general questions, (2) management and leadership, (3) marketing, (4) finance, and (5) operations and was answered by the relevant top managers of the companies.

The survey was carried out between October 2018 and July 2019 and administered by TÁRKI, a company specialized in empirical social science research. Overall, 2062 companies with at least 50 employees were contacted and 234 participated in the survey. In this chapter, we use data from parts 2 and 5 of SMEs in the process industry (97 companies), 63 in the size category of 50–99 employees, and 34 with 100–249 employees.

For this analysis, we chose some specific questions, which can be related to the barriers, drivers and opportunities of Industry 4.0.

5.5 INDUSTRY 4.0 BARRIERS

As a novel form of innovation, Industry 4.0 projects encounter similar barriers than other types of innovation. Although the questionnaire did not explicitly examine Industry 4.0 barriers, we believe that the question on general barriers to innovation may indicate the factors that hold back Industry 4.0 projects. This belief is supported by the fact that during the investigation period (2016–2018), the innovation challenges the SMEs in the manufacturing industry encountered were strongly related to Industry 4.0.

As shown in Figure 5.2, the most important barrier to innovation was the high costs of innovation, rated 3.46 on average, on a 1 (negligible barrier) to 5 (decisive barrier) scale. The shortage in the skilled workforce is a decisive barrier for 27 percent of the sample. The second most frequently mentioned (22 percent) decisive challenge was related to innovation management. These three obstacles are followed by a series of challenges that are relevant for 50–60 percent of the companies. These barriers are the missing know-how for the new technologies, the conservative attitude of the customers, the effect of the large companies, followed by the lacking external or internal financial resources.

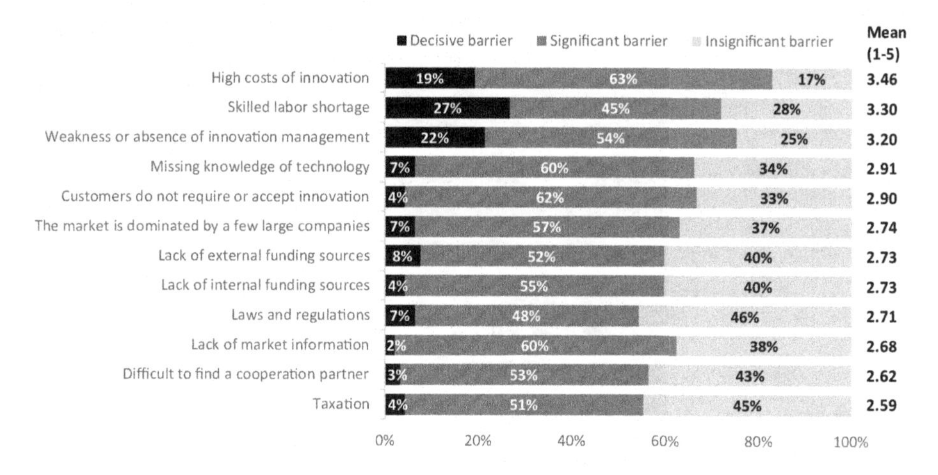

FIGURE 5.2 Barriers to innovation based on our research.

The general barriers to innovation can be completed with some challenges specific to Industry 4.0. Regarding those barriers identified by Horváth and Szabó (2019), the Hungarian SMEs may encounter the lack of technology integration as an obstacle. 37 percent of the respondent companies do not even deploy an ERP system. 26 percent of the managers admitted the lack of planning and frameworks of Industry 4.0 projects. Planning and management, as well as the IT infrastructure-related skills, appeared at the bottom of the list when the companies ranked their company's competences.

5.6 INDUSTRY 4.0 DRIVERS AND ENABLERS

The survey requested the respondents to assess the potential of Industry 4.0. The questionnaire listed eleven benefits in order to evaluate the expectations that drive the current and future Industry 4.0 initiatives. Company executives were asked to choose between three options: significant benefits, moderate benefits, or no benefits expected. The survey revealed the companies' positive attitude towards Industry 4.0. The ratio of the "benefits are not expected" response is lower than 50 percent in all cases. Figure 5.3 ranks the potential benefits based on the survey results. The most expected benefit is making faster decisions (53 percent), followed by the potential for gathering new information (45 percent), increasing the productivity (43 percent), and optimizing the services and the customer relations (39 percent). Surprisingly, cost-cutting and revenue growth appear less attractive: two-third and three-quarter of the respondents, respectively, do not expect significant benefits at these fields.

Meanwhile, the respondents also supported the assumption that Industry 4.0 projects aim to establish new business models (40 percent), new products (35 percent), or to reach new markets and customers (34 percent), or distribution channels (29 percent). A less expected benefit is that Industry 4.0 may result in a more attractive workplace, only 53 percent of the companies foresee any, rather moderate, progress.

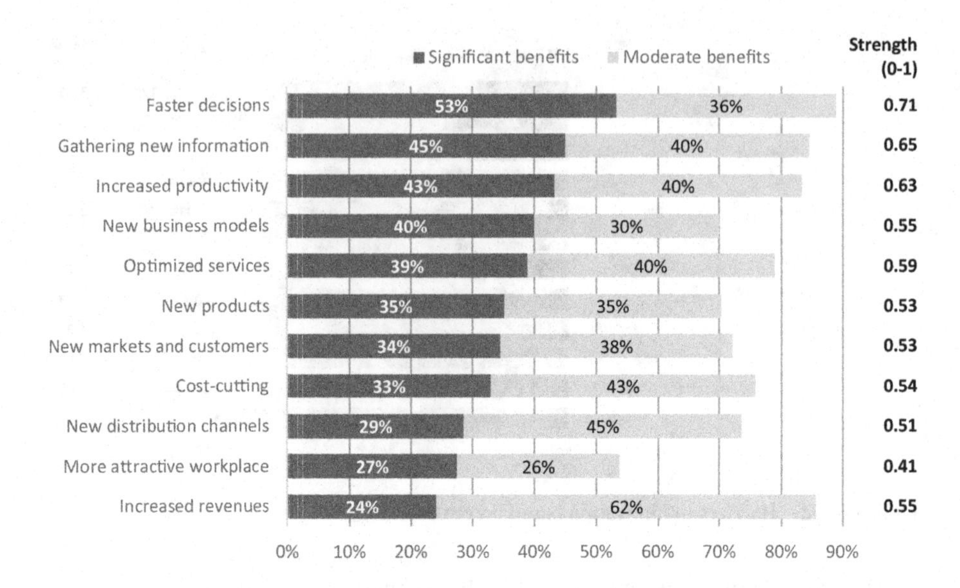

FIGURE 5.3 Expected benefits from Industry 4.0 projects based on our research.

In order to understand why some Hungarian SMEs are more developed in the field of Industry 4.0 than others, we examined the internal drivers of these innovations. The respondents assessed their preparedness in eleven areas, grouped into four categories of Industry 4.0 enablers:

- Top management that understands the digital challenges and opportunities.
- A resource readiness, i.e., the company possesses the necessary knowledge and skills about technology, as well as the financial resources.
- Explicit plans and procedures like a digital strategy and a framework for Industry 4.0 project initiation and implementation.
- Openness and innovative culture that covers the tracking of the state-of-the-art industry innovation, the experimentation with new digital technologies, the opportunity for bottom-up innovation, and the willingness to take risk or openness to change related to Industry 4.0.

Based on these readiness factors, four clusters emerged in the sample of the Hungarian SMEs of the manufacturing industry. Figure 5.4 highlights the differences between these groups. The first group, 22 percent of the sample, covers the most prepared companies. Executives of these companies assessed themselves around the 4.50 score on a 1 to 5 Likert scale (5 meaning full agreement with the readiness criteria). The top category where these companies excel is an existing framework for Industry 4.0 project implementation (4.80). Although the self-assessment may include some bias towards the positive direction, the scores may be analyzed relative to each other. In the most Industry 4.0 prepared group, it is an important finding that openness and innovation culture received the highest score among the four enablers. In contrast

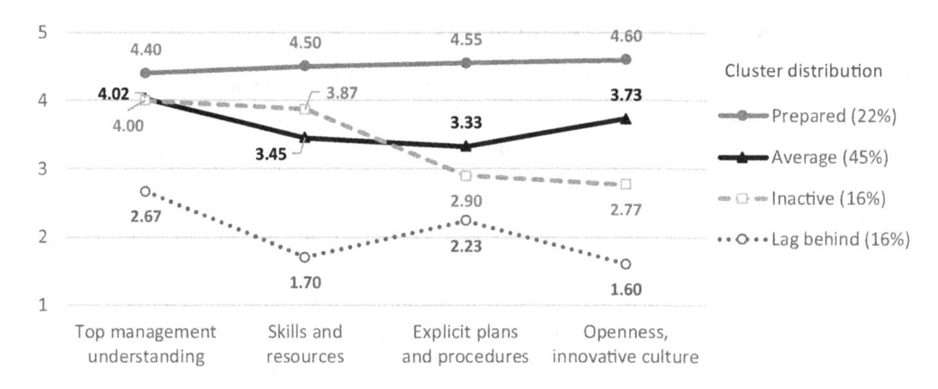

FIGURE 5.4 SME clusters of Industry 4.0 readiness based on our research.

to the other three clusters where respondents reported the highest score for the top management understanding.

The respondents of the second cluster, the largest group (45 percent) in the sample, moderately agreed with their preparedness in almost all Industry 4.0 enablers. Higher scores appeared in the field of top management understanding (4.02) and the enablers related to openness and innovative culture. Skills, resources, and explicit plans are less prevalent at these companies. However, the fact that all readiness criteria received an average score may indicate that the respondents are aware of the importance of these factors but admit that they are not excelling in any of these. The least present enabler is an existing framework for Industry 4.0 project implementation (3.17).

The third cluster labelled as "inactive" (16 percent in the sample) displays a specific pattern. These companies reported that they have the necessary financial resources (3.93), technology knowledge and skills (3.80), and management commitment (4.00) to initiate Industry 4.0 projects. However, they fall behind regarding the direct activities that enable Industry 4.0. An explicit digital strategy is typically missing (2.80), the Industry 4.0 projects do not have a standard process (3.00), they do not track the latest digital developments (3.27) or even experiment with them (2.27).

As the dotted line on Figure 5.4 indicates, 16 percent of the Hungarian SMEs in the sample lag behind significantly. These companies are rather unprepared with scores below 2.00 in the field of the skills and resources, as well as in the openness and innovative culture that enable Industry 4.0. Slightly higher scores emerge for top management understanding (2.67), digital strategy (2.40), and standard implementation processes (2.07), though these companies also lag behind compared to the first three clusters in these areas.

In summary, openness and innovative culture (tracking the latest developments, risk-taking etc.), as well as standardized procedures for Industry 4.0 projects may drive the Industry 4.0 developments, as these readiness criteria turned to be the most important at those companies that are the most prepared for Industry 4.0 in general.

5.7 INDUSTRY 4.0 OPPORTUNITIES

In order to identify the Industry 4.0 opportunities in the Hungarian SME sector, we analyzed the perceived gap between the individual companies and the international leaders of the industry. We examined five areas of Industry 4.0 enhancements. The first three areas are developments in the field of manufacturing operations: modern manufacturing technology, automated operations, and digitalised processes. The second two areas focus on the output, inquiring the smart components of the products, as well as how the products are connected to real-time networks. We asked the production managers to assess how their company progressed in these areas.

Half of the respondents reported modest or moderate progress in more or less all of the five areas. Especially modern technologies are more developed. At the other extreme, 23 percent of the sample made no or very limited progress in these fields. The remaining 27 percent of the companies in the middle showed a rather specific pattern. These companies are more developed in the field of smart and connected products, as well as in the field of automation while lagging behind regarding the modern technologies and the digitalized operations. Modern technologies have an interesting pattern in the sample. This is the Industry 4.0 field where the ratio of more developed companies is the highest (24 percent). On the other hand, almost half of the sample (49 percent) admits that their company lags behind in this field.

We also asked the managers to assess the Industry 4.0 progress of the international leaders of their industry. Around two-thirds of the managers perceives a significant gap between their achievements and the industry leaders. As shown in Figure 5.5, the gap is especially large in the field of digitalized operations, where the average progress of the Hungarian companies (1.3) is around half of the progress of the leaders (2.5). In the case of the automated operations, the Hungarian SMEs rated themselves as more developed (1.6) but also answered that the international companies further

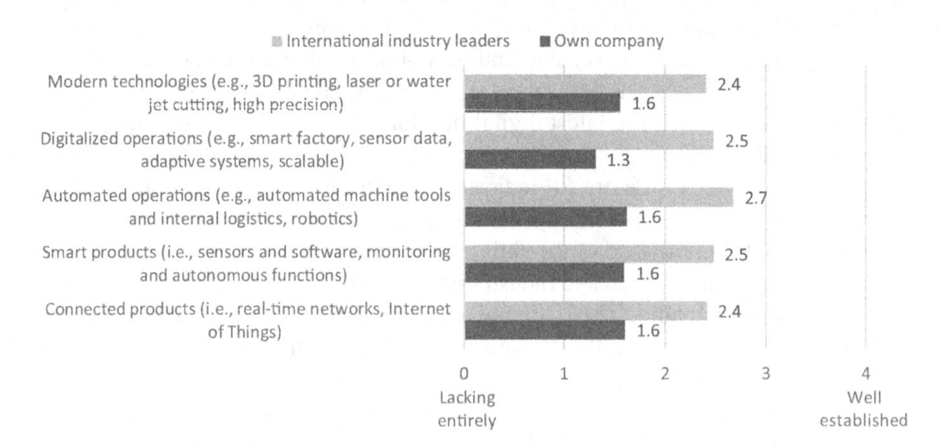

FIGURE 5.5 Penetration of Industry 4.0 enhancements compared to the perceived practice of the international industry leaders based on our research (0=lacking entirely, 4=well established).

excel in this field (2.7). In summary, the perceived gap offers an extensive opportunity to initiate new Industry 4.0 projects. Potential development areas emerge in the field of manufacturing processes (technology, automation, digitalization) and product development (smart and connected products).

5.8 DISCUSSION AND CONCLUSIONS

In Hungary, SMEs, similarly to large companies, are relatively weak in innovation (Hergár, 2020). Although the government makes efforts through various programs and frameworks to intensify innovation activities, and especially, the level of digitalization in the economy, Hungary performs low in digitalization-related international rankings (e.g., European Commission, 2019).

In this chapter, we used data collected from Hungarian manufacturing SMEs to provide an overview of the barriers, drivers, and opportunities of Industry 4.0.

Barriers. We found that high costs, the shortage of skilled labor and the lack of innovation management knowledge are the most important barriers for SMEs. These findings support the results of an interview-based research by Horváth and Szabó (2019), which showed that the most important barriers to Industry 4.0 are the human resources (competences, training), the financial resources (funding, shortcomings in tendering processes, profitability) and the management reality (skills, experience, conscious planning). Szerb, Komlósi and Páger (2020) also listed these factors among the barriers.

Drivers. The respondents' enthusiastic expectations regarding the information gathering, the decision-making, and the new business models may indicate that they have limited experience with realized Industry 4.0 projects since cost-cutting and revenue growth are top expectations in the literature (Juhász, 2018; Moeuf et al., 2020). However, it can indicate that the Hungarian SMEs consider those potential direct benefits as the sources of cost-cutting and revenue growth. Our clusters on Industry 4.0 readiness resonate quite well with the groups identified by Müller, Buliga and Voigt, 2018). In their interview-based research they also identified four groups based on the progress companies made towards Industry 4.0: (1) craft manufacturers (do nothing, 38 percent of companies); (2) preliminary stage planners (have plans, 9 percent); (3) Industry 4.0 users (29 percent); and (4) full-scale adopters (24 percent). Our analysis gives more insight into this issue, showing that managerial understanding is the first step to start into the direction of Industry 4.0. Furthermore, openness and innovation culture seem to be important enablers of successful digital transformations.

Opportunities. There is a large perceived gap between leading international companies and Hungarian SMEs in every investigated aspect of Industry 4.0. The gap can be partly explained by the fact that SMEs were surveyed while the industry leaders are most probably well-known large companies, and large companies are ahead elsewhere, as well (Szász et al., 2020). Meanwhile, we also found that, regarding the application of modern technologies and automated operations, some respondents – 6 and 5 percent, respectively – placed themselves ahead of the international industry leaders. This result can have at least two explanations: first, due to governmental support there is a limited group of reference factories and experimental plants in

Hungary. Also there is an innovative group of SMEs, who can indeed be ahead of the competition (Hergár, 2020). Second, our general experience of human nature, that the least someone knows a topic, the more s(he) thinks about her/himself. So less knowledgeable managers can give overoptimistic answers about the status of their companies.

One important limitation of our study is that the focus of the questionnaire was competitiveness, and not Industry 4.0 itself, even though digitalization was one of the key issues.

5.9 ACKNOWLEDGEMENT

The data collection was managed by TÁRKI Zrt. between October 2018 and July 2019. The support for data collection was provided by OTP Nyrt. and Vállalatgazdaságtan Tudományos és Oktatási Alapítvány.

REFERENCES

Abonyi, J., Czvetkó, T. and Honti, G. (2020) *Are Regions Prepared for Industry 4.0? – The Industry 4.0+ Indicator System for Assessment*. Cham: Springer Nature Switzerland AG 2020. doi: 10.1007/978-3-030-53103-4.

Autio, E., Szerb, L., Komlósi, É. and Tiszberger, M. (2019) *EIDES 2019 The European Index of Digital Entrepreneurship Systems, Publications Office of the European Union*. Luxembourg. doi: 10.2760/107900.

Borbás, L. (2014) 'Supporting SMEs in Central-Eastern Europe', in *Management, Enterprise and Benchmarking – In the 21ST Century*. Budapest: Óbuda University Keleti Károly Faculty of Business and Management, pp. 87–106.

Chikán, A. (2008) 'National and Firm Competitiveness: A General Research Model', *Competitiveness Review*, 18(1–2), pp. 20–28. doi: 10.1108/10595420810874583.

EC (2019) *SBA Fact Sheet 2019 Hungary*. Brussels: European Commission.

EC-ESIF (2020) *Competitiveness of SMEs, European Structural and Investment Fund*. Available at: https://cohesiondata.ec.europa.eu/themes/3#.

Éltető, A. (2019) *Effects of Industry 4.0 on Reshoring Investments – Hungarian Experiences, Working Paper 251*. Budapest. Available at: https://eprints.lib.hokudai.ac.jp/dspace/bitstream/2115/39616/1/JESW15_003.pdf.

Essakly, A., Wichmann, M. and Spengler, T. S. (2019) 'A Reference Framework for the Holistic Evaluation of Industry 4.0 Solutions for Small- and Medium-Sized Enterprises', *IFAC-PapersOnLine*. Elsevier Ltd, 52(13), pp. 427–432. doi: 10.1016/j.ifacol.2019.11.093.

European Commission (2019) *Digital Economy and Society Index (DESI) 2020: Thematic Chapters, European Commission*. Available at: https://ec.europa.eu/digital-single-market/en/desi.

Götz, M., Éltető, A., Sass, M., Vlčková, J., Zacharová, A., Ferencikova, S. and Kaczkowska-Serafińska, M. (2020) *Effects of Industry 4.0 on FDI in the Visegrád countries, Final Report of the Project No. 21920068 Financed by the International Visegrad Fund*. Vistula University, the Department of Business and International Relations. Available at: https://industry40fdi.com/2020/11/02/final-report-effects-of-industry-4-0-on-fdi-in-the-visegrad-countries/

Halmosi, P. (2020) 'The Interpretation of Industry 4.0 by Hungarian Technology-Oriented Startups', *Timisoara Journal of Economics and Business*, 12(2), pp. 149–164. doi: 10.2478/tjeb-2019-0008.

Hergár, E. (2020) *Versenyképessegi jelentés 2020 (Competitiveness Report 2020)*. Budapest. Available at: www.mnb.hu/letoltes/versenykepessegi-jelentes-hun-2020-0724.pdf.

Holicza, P. and Tokody, D. (2016) 'Field of Challenges: A Critical Analysis of the Hungarian SME Sector Within the European Economy', *Hadmernok*, 3(September), pp. 110–120. Available at: http://hadmernok.hu/163_09_holicza.pdf.

Hope, K. (ed.) (2019) *Annual Report on European SMEs 2018/2019*. Luxembourg: European Commission – EASME. doi: 10.2826/500457.

Horváth, D. and Szabó, R. Z. (2019) 'Driving Forces and Barriers of Industry 4.0: Do Multinational and Small and Medium-Sized Companies Have Equal Opportunities?' *Technological Forecasting and Social Change*. Elsevier, 146(October 2018), pp. 119–132. doi: 10.1016/j.techfore.2019.05.021.

Hungarian Insider (2020a) *A New Digital Framework Strategy is Being Developed, Hungarian Insider*, June 16, 2020. Available at: https://hungarianinsider.com/a-new-digital-framework-strategy-is-being-developed-4543/

Hungarian Insider (2020b) *Irinyi Plan Will Continue in 2020, Hungarian Insider*, January 14, 2020. Available at: https://hungarianinsider.com/irinyi-plan-will-continue-in-2020-3472/

IMD (2020) *IMD World Digital Competitiveness Ranking 2019, IMD World Competitiveness Center*. Lausanne. Available at: www.imd.org/globalassets/wcc/docs/release-2017/world_digital_competitiveness_yearbook_2017.pdf.

Istrate, D. (2019) *Hungary Reveals New SME Strategy, Hungarian Chamber of Commerce and Industry*, November 8, 2019. Available at: https://mkik.hu/en/MORE%20NEWS/hungary-reveals-new-sme-strategy

ITU (2017) *Measuring the Information Society Report, Itu*. Geneva: International Telecommunication Union (ITU).

IVSZ and Microsoft (2019) *A digitális gazdaság súlya a magyar nemzetgazdaságban (The weight of the digital economy in the Hungarian national economy)*. Budapest. Available at: https://ivsz.hu/a-digitalis-gazdasag-sulya-2019/.

Juhász, L. (2018) 'Overview of Industry 4.0 Tools for Cost-Benefit Analysis', *Tér Gazdaság Ember*, 4(6), pp. 51–71.

Megyeri, E. and Somosi, S. (2019) 'Opportunities and Obstacles for Hungarian Economic Players Along the Roads Being Paved By 4th Industrial Revolution', *International Scientific Journal 'Industry 4.0'*, IV(5), pp. 246–249.

Moeuf, A., Pellerin, R., Lamouri, S., Tamayo-Giraldo, S. and Barbaray, R. (2020) 'Identification of Critical Success Factors, Risks and Opportunities of Industry 4.0 in SMEs', *International Journal of Production Research*. Taylor & Francis, 58(5), pp. 1384–1400. doi: 10.1080/00207543.2019.1636323.

Müller, J. M., Buliga, O. and Voigt, K. I. (2018) 'Fortune Favors the Prepared: How SMEs Approach Business Model Innovations in Industry 4.0', *Technological Forecasting and Social Change*. Elsevier, 132(December 2017), pp. 2–17. doi: 10.1016/j.techfore.2017.12.019.

Nagy J., Oláh, J., Erdei, E., Máté, D. and Popp, J. (2018) 'The Role and Impact of Industry 4.0 and the Internet of Things on the Business Strategy of the Value Chain-the Case of Hungary', *Sustainability (Switzerland)*, 10(10). doi: 10.3390/su10103491.

Schwab, K. (ed.) (2019) *The Global Competitiveness Report 2019, World Economic Forum*. Geneva: World Economic Forum.

Szalavetz, A. (2018) 'Chronicle of a Revolution Foretold – in Hungary Industry 4.0 Technologies and Manufacturing Subsidiaries', *Jstudies in International Economics : Special Issue of Kűlgazdaság*, 2(2), pp. 29–51. Available at: http://real.mtak.hu/77885/.

Szalavetz, A. (2020) 'Digital Transformation – Enabling Factory Economy Actors' Entrepreneurial Integration in Global Value Chains?' *Post-Communist Economies*. Routledge, 32(6), pp. 771–792. doi: 10.1080/14631377.2020.1722588.

Szász, L., Demeter, K., Rácz, B. G. and Losonci, D. (2020) 'Industry 4.0: A Review and Analysis of Contingency and Performance Effects', *Journal of Manufacturing Technology Management*, ahead-of-print(ahead-of-print), p. 28. doi: 10.1108/JMTM-10-2019-0371.

Szerb, L., Komlósi, É. and Páger, B. (2020) 'Új technológiai cégek az ipar 4.0 küszöbén – A magyar digitális vállalkozási ökoszisztéma szakértői értékelése (New Tech Firms in the Era of Industry 4.0 – Expert Survey of the Hungarian Digital Entrepreneurial Ecosystem)', *Vezetéstudomány/Budapest Management Review*, LI(6), pp. 81–96. doi: 10.14267/ VEZTUD.2020.06.08.

Telekom (2020) *A vállalkozások több mint fele digitális fejlesztésre készül a járvány után (More than half of businesses are preparing for digital development after the epidemic)*. Budapest. Available at: www.telekom.hu/rolunk/sajtoszoba/sajtokozlemenyek/2020/junius_19.

van der Linden, N., Enzerink, S., Geilleit, R., Dogger, J., Claps, M., Wennerholm-Caslavska, T., Mbacke, M., Pallaro, F., Noci, G., Benedetti, M., Marchio, G. and Tangi, L. (2020) *eGovernment Benchmark 2020: eGovernment That Works for the People | Shaping Europe's Digital Future, European Commission*. Brussels. Available at: https://digital-strategy.ec.europa.eu/en/library/egovernment-benchmark-2020-egovernment-works-people.

Zhu, J., Aquaro, V., Alberti, A., Daljani, E., Eren, Y. E., Gatan, D., Kauzya, J-M., Korekyan, A., Kwok, W. M., Le Blanc, D., Losch, M., Purcell, R., Reci, E., Susar, D. (2020) *E-Government Survey 2020 – Digital Government in the Decade of Action for Sustainable Development*. New York. Available at: https://publicadministration.un.org/egovkb/Portals/egovkb/Documents/un/2020-Survey/2020 UN E-Government Survey (Full Report).pdf.

6 The Adoption of Industry 4.0 Technologies

The Case of Italian Manufacturing SMEs

Rubina Romanello and Maria Chiarvesio

CONTENTS

6.1 INTRODUCTION

The world has entered its fourth Industrial Revolution with the introduction and spread of the industrial Internet of Things (IoT) and other related technologies (Kagerman et al., 2013; Liao et al., 2017). The digital transformation is a worldwide phenomenon impacting societies, communities, organizations, and companies. Governments, policy makers, practitioners, and academics have widely acknowledged the potential impacts of the adoption of Industry 4.0 (I4.0) technologies around the world. Great expectations have been especially expressed for I4.0's impacts on manufacturing. Particularly, several governments from the old continent have underlined that a widespread adoption of I4.0 technologies could decrease the costs of production and at the same time increase the competitiveness of the manufacturing base in Europe, leading to a "Manufacturing Renaissance" (Mosconi, 2015). In 2011, the German government identified a group of technologies having the potential to shape the future of the country's manufacturing industry (Kagermann et al., 2013). Germany developed a long-term project called "Industrie 4.0" aimed at ensuring the survival of existing manufacturing systems in the long run (Kagermann et al., 2013). However, policy makers and consultants have underlined that this gradual process should be controlled and carefully guided. After this, in fact, other member states

DOI: 10.1201/9781003165880-6

and advanced countries (e.g., Italy, Japan, the US) have developed measures and policies to encompass the digitalization of production processes based on devices autonomously communicating with each other along the value chain (i.e., Smit et al., 2016; Probst et al., 2017), underlining potential impacts both at micro and macro levels (Wee et al., 2016), ranging from individuals to organizations and from industries to societies (Porter & Heppelmann, 2014, 2015).

Born in the policy context, I4.0 has gained momentum around the world. In spite of this, a unique, widely accepted definition is lacking in the literature (Liao et al., 2017; Lu, 2017), even though in the manufacturing context it generally underpins a group of technologies that can facilitate inter-connection and computerization of traditional industry (Lu, 2017). Under this umbrella, Rüßmann et al. (2015) have suggested that this concept is based on nine technological pillars, namely additive manufacturing (3D printing), simulation, horizontal and vertical system integration, the industrial internet of things, the cloud, cybersecurity, robotics and augmented reality (Rüßmann et al., 2015). Each technology has different possible utilization modes, applications and functions related to desired impacts. In fact, manufacturers can implement these technologies to achieve goals in terms of increased productivity, high flexibility, reduced lead times, mass customization through small batch size production, cost reduction, high quality, or increased turnover in terms of opening new markets (Wee et al., 2016; Sauter et al., 2015; Müller et al., 2018). More broadly, Lu (2017) has suggested that I4.0 refers to manufacturing processes that are integrated, adapted, optimized, service-oriented, and interoperable, and correlated with algorithms, big data, and high technologies.

Since 2016, Italy has launched its program to boost first digitalization and then I4.0 adoption through the so-called "Piano Industria 4.0" or "Legge Calenda." In line with this, regional administrations have created policies and incentives to support digitalization and the adoption of I4.0 among all-sized companies. Despite the general interest for I4.0, there is still limited knowledge on the awareness and state of the adoption of I4.0 technologies among manufacturing SMEs. Since technology evolves quickly, small organizations often have to face new technological changes, and the recent evolution of I4.0 is supposed to pose new organizational challenges in this sense. SMEs approaching I4.0 can encounter difficulties, resource and skills constraints in the process (Kleindienst & Ramsauer, 2015; Sommer, 2015). A first aspect relates to low levels of awareness on I4.0 among SMEs, the lack of expertise (Moeuf et al., 2020) and the fact that many firms can have a negative perception of this paradigm (Sommer, 2015). Other barriers relate to privacy and data security issues, the absence of regulations, and the low level of maturity of technologies (Sommer, 2015). Another problematic issue concerns the lack of internal staff qualifications and the general absence or difficulty to shape ICT skilled employees (Sommer, 2015; Horváth & Szabó, 2019). Also, each technology can be applied to different activities of the value chain, leading to completely different impacts (Chiarvesio & Romanello, 2018). For these reasons, the process through which SMEs select and adopt I4.0 technologies may be influenced by a range of internal and external factors. Despite the widespread debate on these themes, empirical research on companies adopting I4.0 technologies is increasing but scarce (Frank et al., 2019). All this makes it interesting to understand how I4.0 is operationally carried out by SMEs, and its potential influence on innovation activities of companies. This chapter investigates this aspect, by highlighting barriers, drivers and opportunities stemming from I4.0. To this purpose, we developed a case-based studybased on in-depth interviews with managers and entrepreneurs of 18 manufacturing SMEs located in Italy.

This study contributes to the still limited literature on I4.0 from a managerial perspective. We contribute to management research by highlighting two main trajectories followed by manufacturing SMEs in adopting I4.0 technologies, particularly showing different applications to business functions and suggesting potential implications in terms of the companies' innovation activities. In fact, most existing studies on I4.0 belong to the fields of engineering and computer systems (Liao et al., 2017). However, reflections from the managerial side are desirable in light of the increasing importance of the topic.

The chapter includes seven sections, including this one. The next section describes the country background and digitalization policies of Italy. The third section describes the methodology. The following sections, respectively, illustrate I4.0 barriers, drivers, and opportunities. Conclusions follow in the last section.

6.2 COUNTRY BACKGROUND AND DIGITALIZATION POLICY IN ITALY

Italy has a long entrepreneurship tradition, with a strong manufacturing base. The digital transformation has involved manufacturing companies, with a predominance of adoption of I4.0 technologies in the metals and machinery sector (Centro Studi Confindustria, 2019). In the Italian context, the government reports show that 8.4 percent of companies have adopted at least one technology belonging to I4.0, with this propensity increasing with companies' size (18.4 percent of small companies with at least 10 employees) (Ministero dello Sviluppo Economico – MISE, 2018). However, the report of the Italian Ministry of Economic Development (MISE) (2018) clearly underlines that small companies in Italy have a lower propensity to adopt I4.0 technologies. A recent report by the Italian Institute of Statistics (ISTAT, 2020) is in line with those data: among companies with at least 10 employees, in 2020, the 82 percent of firms has adopted less than 6 among the 12 digital technologies considered, in a basket including I4.0 technologies, but also infrastructural and connectivity solutions, such as management software and or broadband and cloud); however, only the 8 percent has adopted at least two smart products or interconnected machines, robotics or big data analytics, whereas only the 4,5% uses 3D printing solutions in production.

Data can provide different interpretations when considering size and industry. As a result, at the national level, a report investigating the mechanical sector has shown that about 70 percent of companies have already adopted at least two I4.0 technologies in 2015 (Federmeccanica, 2016).

Indeed, the adoption of I4.0 technologies are at least partially a result of the Italian I4.0 national policy plan released in 2016 and aimed at boosting the investments in I4.0 by leveraging a bundle of fiscal incentives, venture capital incentives, ultrawideband spreading, I4.0 training and education support, and the commitment of institutions to increase the awareness about new technologies and their potentialities. The

TABLE 6.1
Main Guidelines of the Italian I4.0 Plan

Super amortization plans	New "Sabbadini" Law	Research and experimentation tax credit	Patent Box	Start-up and innovative SMEs	Guarantuee Fund	Digital Hub and Competence Center
Incentives for investments in software and IT systems supporting the digital and technological transformation of organizations and manufacturing processes.	Support firms requiring financial and banking support to buy new instruments and machinery, plants, tooling for manufacturing and digital technologies (hardware e software).	Tax credit (50%) on incremental investments in R&D.	Preferential taxation on income from intangible assets, including patents, brands, design, know how, software if protected by copyright.	Preferential taxation, bureaucracy simplification, insolvency law and job market facilitated.	Public guarantees (up to 80%) to support firms and practitioners with difficulties of access to credits at favorable rates.	Digital Hubs are new local entities aimed at supporting the digital transformation of companies. Competence centers are referred to specific universities (or pools of universities) to facilitate and intensify research – industry relationships.

Source: Authors' elaboration

plan aimed at introducing more than 10 billion euros in private investments, 11.3 billion euros in private investments in R&D and innovation on I4.0 technologies and 2.6 billion euros in early stage private investments (*www.mise.gov.it/images/stories/documenti/guida_industria_40.pdf*).

Table 6.1 summarizes the guidelines of the Italian I4.0 plan, which includes direct and indirect incentives for investments in I4.0 technologies, preferential taxation policies and bureaucracy simplifications, public guarantees, and the creation of ecosystems in support of I4.0 adoption and spread.

6.3 METHODOLOGY

The aim of the research is to investigate opportunities, drivers, and barriers related to the adoption of I4.0 technologies among a sample of SMEs. In respect of the novelty of the topic, we chose an inductive qualitative research with theory-building purposes (Eisenhardt, 1989; Welch et al., 2011). Exploratory case study research is considered suitable to generate theoretical propositions upon which base future large-scale quantitative testing (Welch et al., 2011). This approach can provide insights into "why" and "how" relationships occur in a particular phenomenon (Eisenhardt, 1989; Welch et al., 2011) and can reveal mechanisms that link different phenomena together (Perren & Ram, 2004). We chose the firm as the "unit of analysis" (Perren & Ram, 2004). The cross-case analysis was carried on inductively, by focusing on "how" and "why" questions.

The sample consisted of 18 manufacturing SMEs in the metals and machinery sector located in the Italian region Friuli Venezia Giulia (FVG). The region has a tradition related to the metals and machinery district, and still hosts the metals and machinery regional cluster. This is a receptive sector to technological advances and, thus, interesting to investigate I4.0. We adopted a purposive sampling approach (Miles & Huberman, 1994) to identify companies in the sector that had already approached I4.0 technologies. The chosen firms had to fulfil to the following criteria: (1) respond to the European definition of SME in terms of turnover (less than 50 million Euro) and staff headcount (less than 250 employees) (EU Recommendation 2003/361, 2003), (2) belong to the metals and machinery sector, (3) be located in FVG region, (4) have adopted at least one of the nine I4.0 technologies described by Rüßmann et al. (2015), namely the IoT, augmented reality, big data and analytics (BDA), 3D printing, horizontal and vertical integration, cloud, simulation, robotics, and cybersecurity.

In 2017 and 2018, we conducted face-to-face in-depth interviews, based on a previously developed semi-structured questionnaire (Miles & Huberman, 1994), with entrepreneurs, CEOs and/or operation/production managers of 18 SMEs. Interviews lasted between 90 to 210 minutes, for a total of about 45 hours. Most interviews were audio-recorded and literally transcribed. Besides, we visited production factories in order to see the applications of I4.0 technologies *in loco*. We also collected and analyzed press and archival data for triangulation purposes. To support cross-case analysis, data were organized in excel tables. Table 6.2 describes the main features of sampled companies. In terms of product and processes, 11 companies manufacture machinery, plants, and cars, whereas seven firms are precision mechanics and

TABLE 6.2
Profile of Sampled Companies

Company	Product	Age	Turnover 2016 (M €)	FSTS (%)	Industry diversification	I4.0 technologies adopted
C1	Packing machines	72	20	90	Medium-Low	IoT
C2	Coffee machines	90	22	50	Low	IoT, 3D printing
C3	Beverage machines and plants	32	10	40	Medium-low	IoT, simulation, horizontal and vertical integration, cloud, BDA, robotics
C4	Programmable ovens	8	6	75	Low	IoT, cloud, BDA
C5	Ecological cars	9	2,5	85	Low	IoT, cloud, BDA
C6	Precision mechanics	38	6,5	65	High	robotics, horizontal and vertical integration, BDA
C7	Mechanical machinery	22	7	55	Medium-high	simulation, vertical integration, the IoT, cloud and BDA
C8	Precision mechanics	40	7,8	90	High	robotics, vertical and horizontal integration
C9	Precision mechanics	50	6	70	High	simulation, robotics, vertical and horizontal integration, cloud
C10	Saws	40	18	80	Low	simulation, vertical integration, robotics, BDA

Company	Product	Age	Turnover 2016 (M €)	FSTS (%)	Industry diversification	I4.0 technologies adopted
C1	Mechanical machinery	14	10	95	Low	IoT, robotics, cloud
C2	Wheeled machinery, cobots	16	2,5	80	Medium-low	Simulation, vertical integration, the IoT, BDA
C3	Trailers	90	10	20	Low	Simulation, vertical integration, the IoT
C4	Mechanical machinery	22	15	85	Medium-high	Simulation, vertical integration
C5	Precision mechanics	28	14	74	Medium-low	Simulation
C6	Machinery and plants	54	20	90	Low	Simulation, the IoT
C7	Precision mechanics	46	3	10	Medium-low	Robotics, the IoT
C8	Precision mechanics	37	8	50	Medium-high	Robotics

Source: Authors' elaboration

components manufacturers. The firm age ranges between 8 and 90 years, whereas turnover ranges between 2.5 and 20 million euros. Most companies have high export shares.

6.4 INDUSTRY 4.0 OPPORTUNITIES

The analysis of the opportunities seized by the companies through the investment in I4.0 technologies has revealed two main patterns of adoption influencing the innovation activities of SMEs. Half of the companies adopted technologies mainly aimed at improving manufacturing processes, in line with process innovation activities. In this sense, companies followed a path that, at its maximum extent, could ideally lead to the smart factory concept. The second group of companies adopted I4.0 technologies in order to mainly achieve product innovation. In particular, the most innovative companies in this group have developed smart and connected products, fully benefiting from the Internet of Things by collecting, analyzing and interpreting data. Figure 6.1 illustrates the number of companies that have adopted each technology, as grouped per production type. Most machinery manufacturers have adopted IoT, simulation, BDA, cloud, and vertical integration, whereas components

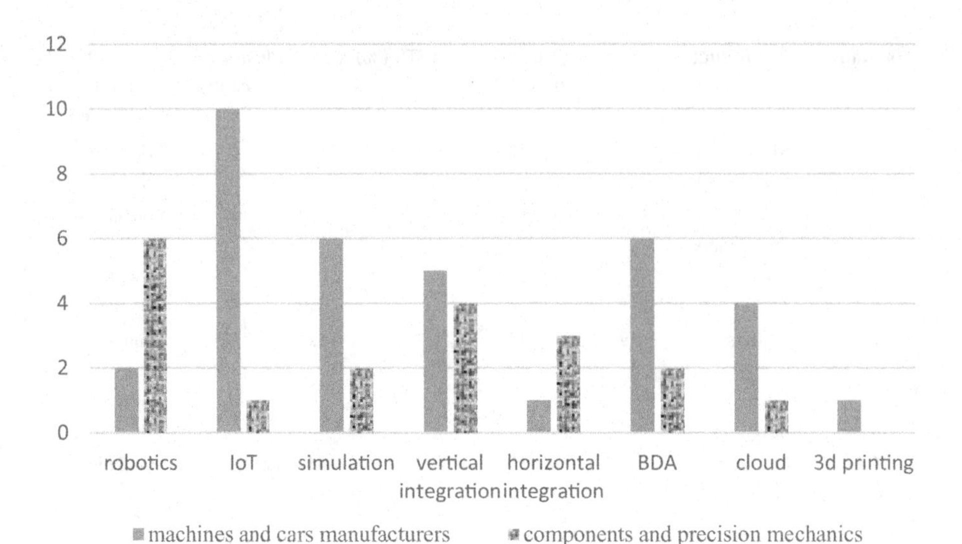

FIGURE 6.1 Number of companies adopting each technology, grouped per product features.

Source: Authors' elaboration

manufacturers tend to adopt robotics, vertical and horizontal integration. Only a few precision mechanics manufacturers use simulation, the IoT, BDA, and cloud.

Figure 6.2 illustrates the number of technologies adopted by the companies as grouped between machinery and cars manufacturers and firms manufacturing components and precision mechanics. As shown by this figure, most machinery manufacturers have adopted more than three technologies, up to a maximum of seven, whereas components and precision mechanics manufacturers provide a different picture: three companies use three technologies, other four firms use three to five technologies.

Table 6.3 shows the clusters among the different technologies. For instance, the IoT is mostly related to BDA (6), simulation (5), cloud (4) and vertical integration (4). Besides the IoT, BDA are related to vertical integration (5) and cloud, simulation, and robotics (4). Beyond this, vertical integration is highly related to simulation (7) and robotics (5).

As far as process innovation, advanced robotics (from autonomous robots to cobots), vertical integration solutions have been adopted to increase productivity, operational efficiency and reduce wastes, both of time and materials. For example, autonomous robots are used to cover the third work shift, as they are able to work autonomously if paired with automated inventories. Simulation has an impact on the quality and effectiveness of manufacturing processes and products. Also, simulation software, when shared with customers, can help in detecting the weaknesses of the production processes (e.g., bottlenecks, breakpoints) and introduce improvements since the designing phase (before prototyping the products for the clients). Besides, simulation opens up the possibility to offer new services to clients. Horizontal and

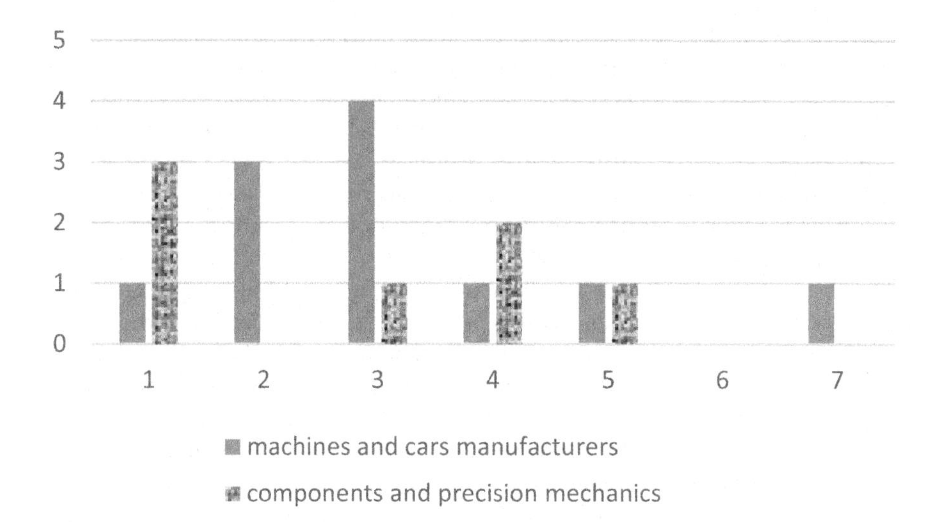

FIGURE 6.2 Number of technologies adopted by the companies split between machines and cars manufacturers and components and precision mechanics manufacturers.

Source: Authors' elaboration

TABLE 6.3
Clusters of Technologies and Total Number of Firms Adopting Each Technology (Diagonal)

Technology	IoT	BDA	Cloud	Vertical Integration	Horizontal Integration	Simulation	Robotics	3D printing
IoT	**11**							
BDA	6	**8**						
Cloud	4	4	**5**					
Vertical integration	4	5	2	**9**				
Horizontal integration	1	2	1	3	**3**			
Simulation	5	4	2	7	4	**9**		
Robotics	2	4	2	5	4	3	**8**	
3D printing	1	0	0	0	0	0	0	**1**

vertical integration are the most impactful I4.0 solutions. Vertical integration allows the interconnection of different functions within the entire organization. However, none of these companies has reached a level of total integration yet, whereas most of them are able to remotely monitor and manage their plants in real-time. More complex is horizontal integration, as it requires the involvement of actors external to the company, such as suppliers, customers and other key value chain partners and requires a certain degree of transparency; on the other hand, horizontal integration may represent an opportunity for small suppliers to strengthen relationships with their clients. Many companies operating in the highly specialized field of mechanical engineering whose clients are mostly large MNEs have already received requests of structuring for horizontal integration. However, whereas some companies have been organizing to implement this application, other firms are against a policy that gives their clients information and rights to impose their priorities on the internal organizational planning and control.

Product innovation through I4.0 mainly consists of developing smart, connected products, which include sensors enabling the data collection and, in some cases, the connection with central systems, other devices and/or the cloud. In the group following this approach, for most companies the creation of smart products has been a natural evolution of product innovation. For example, established machinery manufacturers have continuously invested in product innovation and, particularly, in related technological advances over the years. At the beginning, manufacturers introduced the inclusion of sensors that allowed to extract simple data through the USB pen drive or cables. Now, things have massively changed, with an increasing number of manufacturers producing interconnected machines or entire turnkey plants that can be often remotely real-time monitored and managed.

An important innovation in this domain reflects the increasing attention to service innovation as a component of product innovation towards a servitization logic, following the idea that the service offering should overcome the product selling. Company 3, one of the most extreme cases, still produces machines and plants, while it is transforming its business model to sell service packages, instead of single plants, through a formula that sees customers paying for the use of the machine, rather than for the machine itself. More often, smart connected products allow the offering of new services, such as remote monitoring or predictive maintenance, with strong impacts on the modes of offering post-sale assistance.

6.5 INDUSTRY 4.0 DRIVERS

It has been stated that strategy, not technology, drives digital transformation among organizations (Kane et al., 2015). In line with this statement, the cross-case analysis has revealed that companies have approached I4.0 to achieve different results consistently with the organization's characteristics and market positioning. In particular, the companies selected the technologies that better responded to their necessities related to competitive advantage sources and innovation activities. Indeed, the two patterns of adoption just highlighted that investments were directed to achieve specific objectives. Accordingly, each technology was specifically selected because it was considered suitable and coherent with the company needs and its history.

Our analysis has also allowed the identification of three main categories of drivers, which reflect three respective motivations: cost savings or improvements in performance, market needs, strategic innovation or positioning improvements.

TABLE 6.4
Drivers of I4.0 Investments Categorized by Main Objectives

Market driven investment	Cost driven investment	Strategy innovation
– Product innovation – Product quality – Customer needs – Customer satisfaction – High quality customer pre- and post-sales service – Effective and new value-added services – Customer loyalty enhancement	– Productivity and Efficiency in production – Efficiency and high quality in post-sale service – Performance increases – Quality improvement – Errors prevention	– Service innovation and business model innovation (servitization) – International competitiveness – Integration with customers – lock in effect – Sustainability – Job improvement

Source: Authors' elaboration

Table 6.4 illustrates the specific goals that companies intended to achieve through technological investments made in I4.0, split among market-driven investments, cost-driven investments, and strategy innovation.

As illustrated in Table 6.4, when considering improvements of market relationships, I.40 technologies allow innovating and increasing the product quality and, thus, better respond to customer needs, with positive returns in terms of customer satisfaction. Also, technologies allow firms to offer new services or improve existing ones. As a result, even customer loyalty can be enhanced. The IoT, in a context of product innovation, responds very often to these needs. But even BDA or simulation or process integration are consistent with these goals. For instance, simulation can be used in the design phase to develop customized products that specifically respond to specific needs of client firms, while keeping costs low.

Another driver is related to cost savings and efficiency improvements, both in the case of manufacturing processes and customer services. Technologies adopted to improve product quality, such as simulation or vertical integration, can be included in this category as this aspect not only satisfies the customer, but also reduces and prevents errors and related recovery costs. Typically, automation and process integration are consistent with these objectives. Simulation also allows efficiency improvements in production and more effective products. For example, 3D printing is considered particularly suitable to low-cost manufacture small batches of highly customized plastic workpieces that are needed during the assembly of machines (company 2).

Finally, some companies explicitly addressed I4.0 in order to innovate their business model in line with the servitization perspective. This probably represents the most innovative trend related to digital technologies adoption; however, other strategic objectives can relate to the improvement of the firm's competitiveness or even an explicit attention to sustainability as competitive advantage. In this view, it becomes interesting to consider that some new technologies can be used to improve work conditions and the quality of jobs. For example, this might be the case of automated logistics, which can be used to relieve workers from moving heavy loads.

Our analysis shows potential connections between I4.0 technologies and the drivers: e.g., the IoT leads to product innovation, whereas vertical integration is more inclined to pursue high efficiency objectives. However, the same technologies can be differently exploited by companies depending on the objectives and strategies pursued. For example, simulation can be used both in pre-sales activities, such as the design phase to improve product features and reduce wastes, and in post-sale activities to offer advanced customer services related to the use modes of products after sales. In fact, all the I4.0 technologies can have different functionalities and applications, leading to different objectives and impacts.

6.6 INDUSTRY 4.0 BARRIERS

Our analysis highlights the importance to distinguish between barriers to selection/adoption and barriers to results. As illustrated in Table 6.4, the first group refers to the factors that make it difficult to select the best technologies according to the firm' strategic development plan and to successfully implement the technologies within the company. These barriers include, for instance, the limited knowledge and awareness of I4.0 technologies functionalities and potentialities, the lack of digital competences inside the company, and the lack of sufficient financial resources to plan the investments in I4.0. The national I4.0 plan tries to address the financial issues, stimulating investments in I4.0, and the training and education problem, by favoring the creation and development of I4.0 skills at different levels of the education system, throughout Italy. However, the shortcoming of digital competences, both inside each company and outside its borders, remain a huge problem, which largely prevents companies from investing in I4.0 or – however – slowing down the selection and adoption processes. Another interesting – and relevant – aspect concerns the procedures necessary before digitizing documents, archives, and processes. The digital transformation of companies goes through a series of restructuring and knowledge codification processes that can require long timings – even months – and the involvement of several employees throughout the organization.

Table 6.5 summarizes the barriers preventing companies from adopting I4.0 technologies and the factors impeding the achievement of results (barriers to results).

The next group of barriers, instead, refers to the challenges that companies encounter during the implementation processes, with negative implications on the expected returns on investments. In this category, the most influential aspects relate to data sharing availability with clients and other stakeholders, as the implications in terms of data ownership remain largely unclear. For instance, predictive maintenance requires the access to the client's data, but this is not necessarily well considered by the client firms. In contrast, when this approach is agreed, there are strongly positive implications in terms of performance achievements for the client firm. Still, client firms are not necessarily inclined to pay additional amounts for this new service, which is – instead – taken for granted as a natural consequence of the technological evolution. In the end, technological advances and investments can lead – in some

TABLE 6.5
Barriers to Adoption and Barriers to Results

Barriers to selection/adoption	Barriers to results
– limited knowledge and awareness on I4.0 functionalities, potentialities	– selecting and adopting I4.0 technologies which are not suitable to the firm strategy can embed increased costs and efficiency reduction
– difficulties in understanding which I4.0 technology best suits the firm's manufacturing processes and organizational aspects	– misalignment between I4.0 technologies adopted, and goals expected to be achieved
– lack of the necessary financial resources to carry on a digital strategy including different I4.0 technologies, and fears to invest huge amounts in I4.0 without having the expected returns in short timings	– misleading cost estimates can entail a general increase of costs during the implementation process, postponing the expected benefits due to the adoption of I4.0 technologies
– difficulties in finding capable technological partners	– difficult relationships with technological partners: problems can lead to change partners during the process – Who owns innovations and data?
– restructuring of processes and knowledge codification are necessary before digitizing the processes and archives	– if restructuring processes and knowledge codification processes are not effectively managed, risks of amplifying the redundancies and inefficiencies in the processes
– lack of digital and managerial competences (both internally and externally) necessary to lead the processes restructuring, knowledge codification, the processes of selection/implementation of I4.0, but also to evaluate the proposals of potential technological partners	– the lack of digital competences can lead to huge inefficiencies in the manufacturing systems
– difficult to decide which data could be shared with clients and other stakeholders, and to forecast the potential implications; privacy issues with clients related to the collection and analysis of their data	– fears related to data sharing with clients, which could imply an increase in the pressure of clients on the suppliers' prioritization plans

Source: Authors' elaboration

cases – to offer improved services to client firms without achieving the expected positive returns on the investments made.

6.7 CONCLUSIONS

Our analysis highlights some factors that influence the decisions about the selection and adoption of I4.0 technologies. Moreover, our results show that companies carefully select the I4.0 technologies that are more suitable to the firm strategy, usually following two different trajectories: product innovation or process innovation. As a result, benefits and challenges that stem from the adoption and implementation processes of I4.0 emerged.

Product innovation is usually more related to customer driven investments, whereas process innovation is usually driven by cost savings strategies. However, the reorganization of processes opens the way to renew relationships with customers, particularly, when simulation and horizontal integration are involved. Instead, business model innovation is usually related to both product and process innovation. This opens the way for future researches investigating the relationships existing between I4.0 technologies, business model innovation and servitization strategies, in line with recent research streams that are emerging in management literature (e.g., Müller et al., 2018, 2020; Bortoluzzi et al., 2020).

Our analysis has highlighted some drivers that can stimulate investments in I.40 technologies. Some are oriented to operational efficiency, as they aim at cost and waste reduction and productivity increases. Other drivers are more customer driven: this is the case of I4.0 technologies used to introduce new smart products, or to enlarge the offering by introducing new services. Also, it can be used to increase the customer loyalty and, in some cases, to create lock-in effects due to data sharing with key partners of the value chain. In conclusion, the most interesting experiences are characterized by an overall shift in the business model, which follows strategic drivers of the investments. In fact, some companies used I4.0 technologies to create or improve the conditions to develop a radical servitization strategy, which is a general shift to a new business model where the company sells the usage of products, instead of products themselves.

The capacity to exploit those opportunities should be considered in light of the difficulties that companies can encounter; we found that barriers can hinder the adoption of I4.0, whenever SMEs face financial constraints or lack knowledge needed to identify the right technology or support the implementation process. This is in line with the findings of Horváth and Szabò (2019), who underline that despite good opportunities, SMEs, compared to multinationals, have higher human-resources and financial resource barriers. Beside adoption, other barriers can limit the results, due to poor implementation, wrong strategies, lower than expected response from the market. More in general, our findings support the results of Moeuf et al. (2020), who highlight the lack of expertise as one of the major risks when adopting I4.0 technologies. Indeed, by studying a sample of German industrial firms, Müller et al. (2020) found that limited resources impact on the capacity to explore innovative business models instead of strategies more efficiency oriented.

Companies in the metals and machinery sector are aware of the potentialities of I4.0, even if identifying the best application of each technology consistently with the firm strategy is not that easy or trivial. For this reason, the evaluation, selection, and adoption process of I4.0 can require even some months. Companies are prudent in approaching I4.0, by selecting technologies that are coherent with two main aspects: the firm competitive strategy on the one hand, including the market positioning and customers' features, and the firm innovation strategy on the other hand, intended as the result of the historical approach and evolution of investments in innovation, technology, and R&D activities over the years. As a result, the companies adopted a "cherry picking" approach to I4.0, by selecting and adopting only the technologies that were coherent with firm innovation and competitive strategies. Moreover, technologies are implemented in a creative and various way inside the organizations to serve different purposes. So far, past research has mainly devoted to analyzing the implementation of single technologies in different contexts of application, as done in the case of 3D printing (e.g., Hannibal & Knight, 2018; Laplume et al., 2016), or in different activities of the value chains, by highlighting the different impacts depending on the value chain activity where it was applied. However, we suggest that future researches could give prominence to the interactions existing among the different I4.0 technologies and to the extent to which these interactions were able to impact on the firm's competitive advantage sources and positioning.

6.8 ACKNOWLEDGEMENTS

This research was supported by the region Friuli Venezia Giulia under the projects "The drivers of international competitiveness of manufacturing companies in Industry 4.0" and "Smart and connected products and the competitiveness of companies" under Grants Art. 5, c. 29–33, LR 34/2015.

REFERENCES

Bortoluzzi, G., Chiarvesio, M., Romanello, R., Tabacco, R., & Veglio, V. (2020). Industry 4.0 technologies and the servitization strategy: a good match? *Sinergie Italian Journal of Management*, *38*(1), 55–72.

Centro Studi Confindustria (2019). Dove sta andando l'industria Italiana. 14 Maggio 2019. www.confindustria.it/home/centro-studi/temi-di-ricerca/tendenze-delle-imprese-e-dei-sistemi-industriali/tutti/dettaglio/rapporto-industria+-italiana+-2019.

Chiarvesio, M., & Romanello, R. (2018). "Industry 4.0 technologies and internationalization: insights from Italian companies," In van Tulder, R., Verbeke, A. and Piscitello, L. (Eds.) *International business in the information and digital age (Progress in International Business Research, Vol. 13)*, Bingley: Emerald Publishing Limited, 357–378. https://doi.org/10.1108/S1745-886220180000013015

Eisenhardt, K. M. (1989). Building theories from case study research. *Academy of Management Review*, *14*(4), 532–550.

EU Recommendation 2003/361. (2003). Commission recommendation of 6 May 2003 concerning the definition of micro, small and medium-sized enterprises. https://eur-lex.europa.eu/legal-content/EN/TXT/?uri=CELEX:32003H0361

Federmeccanica. (2016). I risultati dell'indagine Industria 4.0 condotta da Federmeccanica, *Settembre*, 21, 1–56.

Frank, A. G., Dalenogare, L. S., & Ayala, N. F. (2019). Industry 4.0 technologies: Implementation patterns in manufacturing companies. *International Journal of Production Economics*, *210*, 15–26.

Hannibal, M., & Knight, G. (2018). Additive manufacturing and the global factory: disruptive technologies and the location of international business. *International Business Review*, *27*(6), 1116–1127.

Horváth, D., & Szabó, R. Z. (2019). Driving forces and barriers of Industry 4.0: do multinational and small and medium-sized companies have equal opportunities? *Technological Forecasting and Social Change*, *146*, 119–132.

ISTAT (2020). Le imprese usano il web ma solo le grandi integrano tecnologie più avanzate. https://www.istat.it/it/files//2020/12/REPORT-ICT-NELLE-IMPRESE_2019_2020.pdf

Kagermann, H., Wahlster, W., & Helbig, J. (2013). *Recommendations for implementing the strategic initiative industrie 4.0: final report of the industrie 4.0 working group*. Berlin, Germany: Forschungsunion.

Kane, G. C., Palmer, D., Phillips, A. N., Kiron, D., & Buckley, N. (2015). Strategy, not technology, drives digital transformation. *MIT Sloan Management Review and Deloitte University Press*, *14*, 1–25.

Kleindienst, M., & Ramsauer, C. (2015). *SMEs and industry 4.0 – introducing a KPI based procedure model to identify focus areas in manufacturing industry*. 12th Annual International Conference on SMEs, Entrepreneurship and Innovation: Management-Marketing-Economic-Social Aspects.

Laplume, A. O., Petersen, B., & Pearce, J. M. (2016). Global value chains from a 3D printing perspective. *Journal of International Business Studies*, *47*(5), 595–609.

Liao, Y., Deschamps, F., Loures, E. D. F. R., & Ramos, L. F. P. (2017). Past, present and future of industry 4.0-a systematic literature review and research agenda proposal. *International Journal of Production Research*, *55*(12), 3609–3629.

Lu, Y. (2017). Industry 4.0: a survey on technologies, applications and open research issues. *Journal of Industrial Information Integration*, *6*, 1–10.

Miles, M. B., & Huberman, A. M. (1994). *Qualitative data analysis: An expanded sourcebook*. Thousand Oaks, CA: Sage.

Ministero dello Sviluppo Economico – MISE (2018). *La diffusione delle imprese 4.0 e le politiche: evidenze 2017*. Ministero dello Sviluppo Economico. Direzione generale per la politica industriale la competitività e le PMI. Luglio 2018. www.mise.gov.it/images/stories/documenti/Rapporto-MiSE-MetI40.pdf.

Moeuf, A., Lamouri, S., Pellerin, R., Tamayo-Giraldo, S., Tobon-Valencia, E., & Eburdy, R. (2020). Identification of critical success factors, risks and opportunities of Industry 4.0 in SMEs. *International Journal of Production Research*, *58*(5), 1384–1400.

Mosconi, F. (2015). *The new European industrial policy: global competitiveness and the manufacturing renaissance*. London: Routledge.

Müller, J. M., Buliga, O., & Voigt, K. I. (2018). Fortune favors the prepared: how SMEs approach business model innovations in industry 4.0. *Technological Forecasting and Social Change*, *132*, 2–17.

Müller, J. M., Buliga, O., & Voigt, K. I. (2020). The role of absorptive capacity and innovation strategy in the design of industry 4.0 business Models-A comparison between SMEs and large enterprises. *European Management Journal*, doi:10.1016/j.emj.2020.01.002.

Perren, L., & Ram, M. (2004). Case-study method in small business and entrepreneurial research: mapping boundaries and perspectives. *International Small Business Journal*, *22*(1), 83–101.

Porter, M. E., & Heppelmann, J. E. (2014). How smart, connected products are transforming competition. *Harvard Business Review*, *92*(11), 64–88.

Porter, M. E., & Heppelmann, J. E. (2015). How smart, connected products are transforming companies. *Harvard Business Review*, *93*(10), 96–114.

Probst, L., Pedersen, B., Lonkeu, O. K., Martinez-Diaz, C., Novelle Araujo, L., PwC and Demetrius Klitou, Johannes Conrads, Morten Rasmussen, & CARSA (2017). *Digital transformation scoreboard 2017: evidence of positive outcomes and current opportunities for EU businesses*. January 2017. European Commission.

Rüßmann, M., Lorenz, M., Gerbert, P., Waldner, M., Justus, J., Engel, P., & Harnisch, M. (2015). Industry 4.0: the future of productivity and growth in manufacturing industries. *Boston Consulting Group*, 1–9.

Sauter, R., Bode, M., & Kittelberg, D. (2015). *How industry 4.0 is changing how we manage value creation*. Horvàth & Partners, Management Consultants. White Paper, pp. 1–12.

Smit, J., Kreutzer, S., Moeller, C., & Carlberg, M. (2016). *Industry 4.0*. Study for the ITRE Committee, Policy Department A: Economic and Scientific Policy, European Parliament, Brussels.

Sommer, L. (2015). Industrial revolution-industry 4.0: are German manufacturing SMEs the first victims of this revolution? *Journal of Industrial Engineering and Management*, *8*(5), 1512–1532.

Wee, D., Breunig, M., Kelly, R., & Mathis, R. (2016). *Industry 4.0 after the initial hype-Where manufacturers are finding value and how they can best capture it.* McKinsey Digital. https://www.mckinsey.com/~/media/mckinsey/business%20functions/mckinsey%20digital/our%20insights/getting%20the%20most%20out%20of%20industry%204%200/mckinsey_industry_40_2016.ashx

Welch, C., Piekkari, R., Plakoyiannaki, E., & Paavilainen-Mäntymäki, E. (2011). Theorising from case studies: Towards a pluralist future for international business research. *Journal of International Business Studies*, *42*(5), 740–762.

7 Digitalization and the Future of Production

The Case of Baltic States' Manufacturing SMEs

Mantas Vilkas, Morteza Ghobakhloo, and Andrius Grybauskas

CONTENTS

7.1 INTRODUCTION

During the past few years, the ongoing digital industrial transformation, commonly referred to as the fourth Industrial Revolution or Industry 4.0, has radically changed the industrial environment and business dynamics. Under this digital revolution, the last few years have witnessed the emergence of many radically new technologies, such as cloud computing, machine learning, new industrial applications of the internet, and 3D printing (Ghoabakhloo, 2020). Many scholars believe that these new technologies will lead to disruptive change in the coming years, as societies will transition to Industry 4.0 (Brynjolfsson and McAfee 2014; Ford, 2015). At the heart of this transition are ubiquitous digital technologies and connectivity (Iansiti and Lakhani, 2014). Digital technologies allow for the conversion of information from an analog form into a binary form that can be understood by computers in a wide range

DOI: 10.1201/9781003165880-7

of innovations. Digital technologies enable new business models, integrate customer relationship management processes, and facilitate the augmentation and integration of manufacturing processes (Manyika et al., 2012; Schlaepfer et al., 2015).

Companies that have adopted Industry 4.0 technologies have shifted their value propositions away from providing products. They have innovated service business models proposing product and customer support services (Eggert et al., 2011; Lightfoot et al., 2013). Product support services facilitate the usage of products and increase customer satisfaction, while customer support services propose new efficiencies through advanced analytics and algorithms based on the data generated by products and services supporting customers' actions (Lenka et al., 2017; Siggelkow and Terwiesch, 2019). Digital technologies also offer "front-end" digitalization opportunities. The digitalization of front-end customer-relationship management processes allows for new types of customer interaction and self-service opportunities through online product configuration systems (Salvador et al., 2009), online sales platforms (Wang et al., 2016), and digital assistants (Brill et al., 2019). Such opportunities offer companies the possibility to deepen their understanding of customer preferences (Coreynen et al., 2017). Finally, digital technologies facilitate "back-end" digitalization. The digitalization of back-end production processes augments the execution of tasks (Law and Ngai, 2007), integrates production processes and equipment (Hasselbring, 2000), and enables the continuous monitoring of operating status (Ding et al., 2020). Business model innovations, front and back-end digitalization offer companies opportunities that can help them expand customers base, overcome "commodity trap" (d'Aveni, 2010), and increase productivity (Evangelista et al., 2014).

Given the hype surrounding the ongoing digital industrial revolution, the adoption rate of Industry 4.0 digital technologies is progressively increasing (Mittal et al., 2020). Businesses are rushing towards Industry 4.0 with the hope of benefiting from the advantages it provides to early adopters. Industrial reports indicate that industry leaders have been successfully implementing Industry 4.0 digital technologies, developing innovative business models, and sustaining their competitive edge in this hypercompetitive market environment. Nonetheless, Industry 4.0 has appeared to be a significant challenging to smaller businesses (Müller et al., 2020a, Müller et al., 2020b, Müller et al., 2018, Sahi et al., 2020, Raj et al., 2020). Moeuf et al. (2020) argued that the short-term strategy, importance of the SME manager, lack of an expert support function, and short hierarchical line differentiate SMEs from large companies and that SMEs have unique barriers to and opportunities for digitalization. Similarly, Horvath and Szabo (2019) revealed a long list of the drivers of the adoption of digital technologies by SMEs, including many factors such as labor shortages, cost-effectiveness, or pressure from competitors. They also disclosed that companies struggle to adopt digital technologies because of lack of appropriate competences, financial resources, leaders with appropriate experience.

Consistent with European economic development strategies, digitalization has become a strategic priority for SMEs across the Baltic states. However, businesses in these regions have struggled to adapt to the digital economy and implement Industry 4.0 technologies to develop digital business models. The gap in existing knowledge and practice calls for the comprehensive study of the digital transformation of the

SMEs in the Baltic states to enable them to develop digitalization strategies, identify existing risks and challenges, and be aware of possible opportunities. Consistently, the present chapter addresses the issue of the digital manufacturing transformation of Baltic states, primarily through a comprehensive assessment of Lithuanian manufacturing SMEs.

7.2 DIGITALIZATION POLICY IN THE BALTIC STATES

European states are consistently moving towards digitalization and committing more to investments in the digital economy and society. The 2020 Digital Economy and Society Index (DESI) shows that the Baltic states are performing competitively in terms of the digital connectivity and digitization of businesses and public administrations (DESI, 2020a). According to the DESI, Estonia is regarded as a high-performing country, while Latvia and Lithuania are categorized as medium-performing countries among all European countries. Among the three Baltic states, Estonia is ahead in terms of digital transformation by a significant margin, ranking seventh in the digital economy among 28 European states in 2020 (DESI, 2020b). Estonia has comprehensive broadband and internet service development programs, such as ambitious 5G development goals; the country also actively provides the necessary digital business platforms, as prioritized by the "Digital Agenda 2020" strategy. The Estonian government also developed a five-year "2021–2025 Digitalization Strategy" plan at the end of 2020. Despite the country's efforts to enhance its digitalization capacity, Estonian businesses still are not taking full advantage of the opportunities offered for digitalization, which remains a challenge (DESI, 2020b).

Out of 28 European member states, Lithuania ranks 14 in terms of the digitalization of the economy and society (DESI, 2020c). During the past decade, Lithuania has shown a staggering improvement in its integration of digital technologies into public services. Although Lithuania still lags in terms of some digitalization metrics, such as fixed broadband or next-generation access coverage, it is increasingly progressing towards the digital economy. Compared to the European average, Lithuania excels in terms of digital integration, electronic information sharing, and electronic commerce, especially within the SME sector. Lithuania's promising performance in contributing to the digital economy is rooted in its comprehensive digital strategy, such as the 2014–2020 Information Society Development Programme, which has covered some of the more critical requirements of Industry 4.0 digitalization, such as digital skill development, high-speed broadband, cybersecurity, infrastructure reliability, and interoperability. The Lithuanian government has developed and executed various complementary digitalization strategies, such as an interinstitutional action plan, a 2018 national cybersecurity strategy, and a 2019 AI strategy, to facilitate digitalization goals and address the digital divide at both the societal and corporate levels (DESI, 2020c).

Latvia ranks 18th out of the 28 European states in terms of the digitalization of the economy and society (DESI, 2020d), has a high overall connectivity indicator and performs well in the digitalization of public services (DESI, 2020d). Nonetheless, Latvia's lower rank among European countries can be attributed to the its low business digitalization rate, especially among SMEs. The latest digitalization agenda in Latvia dates back to 2013, when the 2014–2020 Information Society Development Guidelines were devised and approved by the government. The 2019–2022 Cybersecurity Strategy and 2014–2020 Science, Technological Development and Innovation Guidelines are among the Latvian government's sectoral strategies to materialize the digital economy, particularly by addressing the gap in terms of digitalization skills (DESI, 2020d).

7.3 COUNTRY PROFILES

The Baltic states share a common history and a close geographical location; Lithuania is bordered by Latvia, and Latvia is bordered by Estonia. Many differences exist that should be noted before any further analysis can take place. The Baltic states are relatively small; in 2020, out of the three countries, Lithuania was the largest, with 2.8 million inhabitants, while Estonia was the smallest, with 1.3 million inhabitants.

Baltic countries' real GDP per-capita is 14,593.3 EUR. In the period from 2000 to 2020, the GDP growth numbers were phenomenal for all three countries, at times reaching over 10 percent growth. According to Eurostat, negative growth was recorded only in 2008 and 2009, and in 2019, sustainable growth levels from 2.1 to 5 percent were still seen (Eurostat, 2020a). Before the emergence of the coronavirus, unemployment levels in these countries were close to the natural unemployment rate. However, labor productivity was relatively low for all Baltic countries. According to Eurostat data on labor productivity per person employed and hours worked, in 2019, Baltic states had a productivity of 75.3, while the EU average was 100 (see Table 7.1).

TABLE 7.1
Key Economic and Manufacturing Indicators of Baltic States

	Lithuania	Estonia	Latvia	Baltic states
Key economic indicators				
Population, millions*	2.8	1.3	1.9	6 (total)
GDP, US$ billions, current*	47.3	23.1	30.3	100.7 (total)
Real GDP per capita, EUR, 2019 (Eurostat, 2020a)	14010	15760	14010	14593.3
Unemployment rate, %*	6.4	6.8	6.5	6.6
Labor productivity, Nominal labor productivity per person employed (Eurostat, 2020b)	78.5	78.8	68.7	75.3
Key manufacturing indicators				
Manufacturing value added, 2010 million US$*	8,352.9	3,435.8	3,015.4	14.8 total
Manufacturing value added in economy, % GDP*	18.4	14.5	10.3	14.4
Manufacturing (NACE C) employment, % working population*	15.2	18.8	13	15.7

	Lithuania	Estonia	Latvia	Baltic states
Manufacturing value-added growth, Annual %*	3.5	2.5	–0.6	1.8
Medium hi-tech & hi-tech industries, % of manufacturing value added*	23.1	28.8	21.5	24.5
CO2 emission per unit of value added, kg/USD*	0.1	0.2	0.3	0.2
Manufacturing labor productivity, Gross value added per person employed, thousand EUR, 2016 (Eurostat, 2020c)	19.6	26.3	18	21.3

* – WEF, 2018

The value added of the manufacturing sector constitutes 14.4 percent of gross value added in the Baltic states. A total of 15.7 percent of employees are working in manufacturing companies. Baltic states enjoy, on average, 1.8 percent annual growth in the value of the manufacturing sector. On average, 15.7 percent of the working population is employed by the manufacturing sector in the Baltic states. The share of the manufacturing sector in the GDP structure is highest in Lithuania, where it constitutes 18.4 percent of the economy. Estonia is the most dependent country in this sector, where 18.8 percent of its working population is employed by manufacturing companies. A total of 24.5 percent of manufacturing value is created by medium-high-tech and high-tech industries. Manufacturing labor productivity constituted 21.3 thousand EUR in 2016, while the EU average manufacturing labor productivity constituted 66.0 thousand EUR. If future growth can be leveraged efficiently with new digital technologies, then Lithuania and Estonia may increase their competitiveness in the long run. Latvia has experienced a declining manufacturing sector, the highest level of CO2 emissions and the smallest number of high-tech companies of these three countries at the time of analysis.

7.4 METHODOLOGY

A literature review, analysis of the primary data collected in Lithuania, and secondary data analysis were used to prepare this chapter. A literature review was used to ground the drivers, barriers, and digitalization opportunities for the manufacturing sector of these Baltic countries. A comparative analysis of Lithuania, Latvia, and Estonia was conducted using secondary data sources such as the World Economic Forum's Readiness for the Future production data (WEF, 2018) and the Digital Economy and Society Index data (DESI, 2020a). The classification of the size of enterprises, as presented in Table 7.2, is used throughout the chapter.

The primary data on the Lithuanian manufacturing sector were collected as part of the European Manufacturing Survey in 2018. The Lithuania case was considered representative of the extent of diffusion of digital technologies in SMEs in the Baltic states. The sampling framework consisted of 6,122 manufacturing sites covering all

TABLE 7.2
Classification of Organizations in the Baltic States

Enterprise classification	Number of Employees
Microenterprise	Less than 10
Small enterprise	10 to 49
Medium enterprise	50 to 249
Big enterprise	249 and more

manufacturing subsectors and represented the country's total population of manufacturing sites. Questionnaire respondents were technical or production managers at manufacturing sites with more than 200 employees and general managers, technical managers, or production managers at manufacturing sites with fewer than 200 employees. A telephone survey was used to collect the data. A stratified random sampling procedure was employed. Strata were defined in terms of four country regions and four size-based classes of organizations (2–19, 20–49, 50–99, 100–249, and 249 and more employees). A total of 2,330 manufacturing sites were contacted. The effective final sample included 500 manufacturing sites, which constitutes a 21.5-percent response rate, and adequately represented the four size classes, regions of the country, and manufacturing subsectors.

7.5 EVALUATION OF THE DRIVERS OF INDUSTRY 4.0 TRANSFORMATION IN THE BALTIC STATES

The possibilities of manufacturing companies competing in the Industry 4.0 context have been evaluated by various methodologies. The World Economic Forum (WEF, 2018), Roland Berger (2014), and McKinsey (Manyika et al., 2012) offered insights into the differences in the manufacturing sector characteristics of different countries. We use the World Economic Forum's Industry 4.0 Readiness Report (WEF, 2018) and supporting sources to describe the manufacturing sector of the Baltic states.

The World Economic Forum's Industry 4.0 Readiness Report focused on 100 countries, representing more than 96 percent of the value added of global manufacturing. The evaluation of a country is based on two dimensions: the structure of production and production drivers. The structure of production considers the current baseline of production. Production drivers represent the evaluation of the presence – or lack – of the key enablers that allow manufacturing companies to capitalize on emerging technologies and transform their production systems. An evaluation of Baltic countries is provided in Table 7.3.

The structure of production is measured by a country's economic complexity and scale of production. Economic complexity reveals the mix of products that a country can manufacture to embed this useful knowledge in the economy. The scale of production measures the current manufacturing value added in the economy. The analysis reveals that Lithuania's current production sector is well positioned, followed by those of Estonia and Latvia.

TABLE 7.3
Evaluation of the Manufacturing Sector of the Baltic States in the Context of Industry 4.0

	Lithuania		Estonia		Latvia	
	Score*	Rank**	Score	Rank	Score	Rank
Structure of Production	5.9	31	5.8	34	4.9	49
Economic complexity	6.8	29	7.4	23	6.5	35
Scale of production	4.5	41	3.3	70	2.5	79
Drivers of Production	5.4	37	6.0	27	5.4	38
Technology & Innovation	4.7	38	5.8	24	4.5	42
Human Capital	5.9	33	6.5	20	5.6	37
Global Trade & Investment	5.0	62	5.8	35	5.6	39
Institutional Framework	6.7	28	7.3	20	6.4	33
Sustainable Resources	7.4	21	6.2	52	8.4	7
Demand Environment	4.0	73	3.9	74	3.4	89

* – 1–10, best; ** – rank out of 100 countries which were included in the evaluation

The analysis of the drivers of transformation that are necessary to successfully compete in the changing manufacturing context reveals a mixed situation. With a population of 6 million people, the Baltic states are characterized by a weak demand environment. The global trade and investment evaluation representing current infrastructure, foreign direct investment, trade facilitation, and market access does not encourage production transformation. The characteristics of technology and innovation, human capital, and institutional framework offer potential drivers that facilitate necessary changes for competition in the Industry 4.0 context.

7.6 THE EXTENT OF THE DIGITALIZATION OF MANUFACTURING COMPANIES: THE CASE OF LITHUANIA

In the previous chapter, we presented a comparative analysis of the Industry 4.0 readiness of the manufacturing companies in the Baltic states. In this chapter, we concentrate on adopting the Industry 4.0 practices of manufacturing organizations using the case of the Lithuanian manufacturing sector. The comparative analysis of the Baltic states according to their Industry 4.0 readiness revealed that the most extensive use of digital technologies for business model innovation and the digitalization of production was found in Estonia (ranked 16th of 100), followed by Lithuania (ranked 27th of 100) and Latvia (ranked 60th of 100) (WEF, 2018). Further, we present Lithuania's case, concentrating on the adoption of digital technologies for business model innovation and front- and back-end digitalization.

Digital technologies enable service innovation in manufacturing companies (Ulaga and Reinartz, 2011; Vilkas et al., 2021). The proliferation of customer and product support services is presented in Table 7.4. Companies provide a varied range

TABLE 7.4
The Provision of Digital-Technology-Enabled Services

	A total pool of 500 enterprises		Microenterprises, %	Small enterprises, %	Medium enterprises, %	Big enterprises, %
	Number of implementors	Implementation frequency, %				
Customer support services						
Full-Service contracts	144	28.8	34.8	25.0	26.6	14.3
Design, consulting, project planning	132	26.4	28.3	25.5	26.6	14.3
Operation of products at customer's site for the customer	70	14.0	14.6	17.8	5.1	NA
Remote monitoring of operating status	48	9.6	11.1	8.2	10.1	7.1
Software development	47	9.4	10.1	7.2	13.9	7.1
Taking over the management of maintenance activities	47	9.4	12.6	6.7	10.1	NA
Renting products, machinery, or equipment	32	6.4	5.6	8.7	2.5	7.1
Data-based services based on big data analysis	8	1.6	NA	1.4	6.3	NA
Product support services (SSPs)						
Installation, start-up	113	22.6	24.7	22.6	20.3	7.1
Maintenance and repair	138	27.6	31.3	26.4	25.3	7.1
Revamping or modernization	80	16.0	18.7	12.5	20.3	7.1
Training	79	15.8	18.2	13.5	19.0	NA
Take-back Services	85	17.0	18.2	14.9	19.0	21.4

of customer support services that support customers' actions. Full-service contracts are offered by 28.8 percent of companies; design, consulting, and project planning are provided by 26.4 percent of companies, and the operation of products at customer sites for the customer are provided by 14.0 percent of companies. The list of services supporting customers' actions ends with services based on big data analysis, which are offered by 1.6 percent of companies. Companies also provide a set of product support services such as the installation, maintenance, repair, revamping, or modernization of products. There are no extensive differences between the diffusion of customer actions' support services among SMEs, microenterprises and large enterprises. These findings are surprising in the context of servitization literature. Large companies have product-related capabilities such as product usage data, product development, manufacturing assets, an experienced product sales force and distribution network, and a field service organization that facilitates service innovation (Ulaga and Reinartz, 2011).

Digital technologies allow for the digitalization of front-end processes that involve customers. Companies use varied technologies that help regarding touchpoints with the customer during the customer experience journey, usually creating self-service opportunities (Table 7.5). Remote support for clients is offered by 33.4 percent of organizations. A total of 15.0 percent of companies offer functional websites, which allow for online training, documentation downloads, and information about error cases. Web services for product configuration are offered by 13.0 percent of organizations. The results reveal that product configuration and functional websites are less prevalent among large companies than among SMEs and micro companies.

Finally, digital technologies offer rich opportunities for back-end digitalization. These opportunities for the digitalization of production systems have received extensive attention from operations and technology management scholars. The prevalence

TABLE 7.5
The Extent of Front-End Digitalization

	A total pool of 500 organizations		Micro, %	Small, %	Medium, %	Big, %
	Number of implementors	Implementation frequency, %				
Remote support for clients	167	33.4	34.8	30.3	38.0	35.7
Online training, documentation, error description	75	15.0	15.7	14.4	17.7	NA
Web services product configuration or product design	65	13.0	16.2	12.5	8.9	NA
Mobile devices for diagnosis, repair, or consultancy	49	9.8	14.1	7.2	6.3	7.1

of digital technologies that enable and integrate the production process is presented in Table 7.6. Software that enables production planning and scheduling, applications allowing for the remote control of facilities and machinery, and digital solutions to provide documentation directly to the shop floor are offered by 43.2 percent, 35.7 percent, and 35.2 percent of organizations, respectively. 3D printing is the least prevalent digital technology, which is exploited by only 4.8 percent of organizations. Large companies tend to deploy back-end digitalization technologies more extensively than do SMEs. The complexity of production systems is higher within large companies compared to small companies. The digitization of complex production systems offers more benefits, thus large companies tend to invest into more into digital technologies comparing to small companies. Large companies also possess financial and human resources that aid in their adoption of back-end digitalization practices.

The Lithuania case reveals the proliferation of digital technology-enabled service innovation and the extent of the digitalization of the front- and back-end processes of manufacturing organizations. The findings reveal that the extent of service

TABLE 7.6
The Extent of Back-End Digitalization

	A total pool of 200 organizations		Small, %	Medium, %	Big, %
	Number of implementors	Implementation frequency, %			
Software for production planning and scheduling (EPR)	86	43.2	33.0	56.8	78.6
Remote controlling of facilities and machinery	71	35.7	34.0	37.7	64.3
Digital solutions to provide documentation directly to the shop floor	70	35.2	34.0	42.7	38.5
Near real-time production control systems (MES)	67	33.7	31.9	37.8	64.3
Digital exchange of product/process data with suppliers/ customers	59	29.6	33.7	24.6	69.2
Systems for automation and management of internal logistics	50	25.1	22.4	26.7	61.5
Simulation of product design and development	48	24.1	19.4	32.9	38.5
Industrial robots	32	14.4	18.2	28.6	21.4
3D printing	10	4.8	5.4	8.5	14.3

innovation and front-end digitalization in SMEs is similar to that in large and micro companies and that SMEs lag in terms of back-end digitalization compared to large companies.

7.7 INDUSTRY 4.0 BARRIERS

Industry 4.0 barriers constitute the factors that impede the obtainment of the capabilities necessary for successful performance in the Industry 4.0 context. A weak demand environment, low R&D expenditures, the extent of the future labor force, and limited access to venture capital constitute important obstacles for the digital transformation of manufacturing companies in the Baltic states.

Weak demand environment. With a population of 6 million people, the Baltic states are characterized by a weak demand environment what may impede investments. Some companies, such as food processing plants, are located close to customers, as such products are difficult to transport. The tendency to move production closer to consumers is gaining traction in many sectors over the world, as lead times have become increasingly crucial for sophisticated buyers. Investors may choose the neighbors of the Baltic states such as Poland, with a population of 37.97 million, for the location of their production facilities. Furthermore, if higher risk and trade barriers are acceptable, then investors may choose other neighbors of the Baltic states such as Russia, with a population of 145 million people, for the location of their production facilities.

Low R&D expenditures. Gross domestic spending on R&D measures the total expenditure on R&D carried out by all resident companies, research institutes, universities, and government laboratories in a country. The Baltic states are characterized by low R&D investment, with Estonia having spent 1.4 percent, Lithuania having spent 1.0 percent, and Latvia having spent 0.6 percent of GDP for research and development in 2018 (OECD, 2020a), while Israel spent 4.9 percent, China spent 2.1 percent, and the US spent 2.8 percent of GDP. The EU average stands at 2 percent. The low expenditures on R&D hamper product and process innovation in the Baltic states. Furthermore, the governments of the Baltic states do not act as innovation catalyzers. The government procurement of advanced technology products is ranked 41st (Estonia), 76th (Lithuania) and 90th (Latvia) out of one hundred countries (WEF, 2018). Low research and development expenditures coupled with passive state positions towards innovation constitute important barriers to the transformation of the manufacturing sector in the Baltic states.

The extent of the future labor force. The future labor force is an important factor characterizing countries' ability to compete in the Industry 4.0 context. A strong employee base provides a labor force for difficult-to-automate manufacturing operations. Product and service innovation and front- and back-end digitalization require talent that possesses advanced skills. In recent decades, the emigration trends in the Baltic states have been disturbing. The joining of the Baltic states in the EU in 2004 was followed by the freedom of movement of Baltic state citizens among EU countries. Hundreds of thousands of skilled and unskilled laborers considered this possibility, especially during the economic recessions in 1998 and 2008. As a result, the population declined by 19.2 percent in Estonia, 23.8 percent in Lithuania and

25.5 percent in Latvia from their peak census figures in the 1990s (Statista, 2020). The situation was exacerbated by the weak ability of these countries to attract and retain talent. The Baltic states' ability to attract talent was ranked 51st (Estonia), 83rd (Lithuania) and 89th (Latvia) of all countries (WEF, 2018). This free movement in the EU positions new members in a disadvantageous position in terms of retaining and attracting talent due to differences in wage levels across countries.

Limited access to venture capital. Venture capital is an important driver of innovation. Venture capital funds increase the availability of finance, allow for the acceptance of more risk in investments (Maier II and Walker, 1987), and contribute to the creation of new technological ecosystems such as cloud ecosystems (Breznitz et al., 2018). Venture capital deals accounted for 459.8 US$ million in Lithuania, 292.8 US$ million in Estonia and 524.8 US$ million in Latvia, and such amounts constitute 10.4, 12.2 and 18.3 venture capital deals (US$/GDP). This ranks the Baltic states in the 53rd (Latvia), 62nd (Estonia) and 68th (Lithuania) positions among one hundred ranked countries (WEF, 2018). A lack of venture capital is in part mitigated by strong governmental support. However, governmental support tends to be slow to identify emerging technologies, includes labor-intensive project management procedures and relies on complicated public tender systems.

Having reviewed the barriers, in the next chapter, we focus on the drivers of digitalization of the manufacturing industry in the Baltic states.

7.8 INDUSTRY 4.0 DRIVERS

Industry 4.0 drivers refer to the socioeconomic factors that facilitate digitalization trends. High exposure to regional and global value chains, well-developed digital infrastructure, the use of digital technologies and the continuous increase in economic complexity drive manufacturing industry transformation in the Baltic states.

High exposure to regional and global value chains. Baltic state manufacturing companies export approximately 65 percent of their production (LSD, 2019). The main export partners of Estonia are Russia (11.7 percent), Germany (10.6 percent) and Finland (8.5 percent). Lithuanian exports to Russia (13.3 percent), Latvia (9.2 percent), and Poland (7.6 percent). The main export destinations of Latvia are Lithuania (14.2 percent), Estonia (9.2 percent), and Russia (8.34 percent). Russia stands as an exception in terms of the top three export partners of the countries. Almost all other exports are oriented to EU countries and the UK. An extensive part of exported goods is produced for further processing by regional and global value chains. For example, Lithuania is among the top five IKEA suppliers. Knowledge spillovers, the transfer of managerial practices, and increased productivity are among the benefits of participation in global value chains (Cheng et al., 2015).

Well-developed digital infrastructure and the use of digital technologies. The Baltic states have a well-developed digital infrastructure. According to measures of mobile-cellular telephone subscriptions, LTE mobile network coverage, and Internet users, these countries are among the thirty best-ranked countries (WEF, 2018). Estonian and Lithuanian companies' CEOs positively perceive firm-level technology absorption and the use of digital-technology-enabled business models. The executive opinion survey of CEOs in Estonia and Lithuania reveals an optimistic evaluation of the extent of the use of digital technologies and digital-technology-enabled business models, positioning these counties in the first quartile of the World Economic Forum's Industry 4.0 Readiness evaluation (WEF, 2018). The perception of the use of digital technologies and their enabled business models in Latvia is less optimistic, placing the country in the sixty-third position among one hundred ranked countries. The case of Lithuania presented here confirms that manufacturing companies extensively use service-oriented business models and digitalize front- and back-end operations.

Increasing economic complexity. A country's economic complexity is a measure based on the diversity of the exports it produces and the number of other countries able to produce them (Observatory of economic complexity, 2020). High economic complexity is associated with a diversity of knowledge in a country. The economic complexity of the Baltic states is ranked in the first one-third of industrialized counties (WEF, 2018). Furthermore, on average, 24.5 percent of manufacturing value is created by medium-high-tech and high-tech industries. Such numbers may seem low compared to industry leaders such as Germany, in which medium-high-tech and high-tech industries contribute 61.4 percent of the manufacturing value added. However, the current economic complexity and medium-high-tech and high-tech industries have been mainly achieved during the last thirty years.

In the next chapter, we concentrate on the opportunities that may help mitigate the barriers to and speed up the digitalization of manufacturing companies.

7.9 INDUSTRY 4.0 OPPORTUNITIES

Industry 4.0 opportunities are favorable developments that, if seized, may become factors facilitating the development of the manufacturing industry. The pandemic and Brexit have offered the possibility for countries to regain immigrants and build a future labor force. The capitalization of digital skills, following Estonia, would provide the necessary traction for the region to facilitate innovation in manufacturing.

The COVID-19 pandemic and Brexit. Population decline and an aging society are among the significant barriers that distract foreign investments and limit the growth of existing manufacturing companies. Increasing average gross annual earnings and recent geopolitical shifts constitute an opportunity to build a future labor force. The estimated labor hour costs are still relatively lower in the Baltic states than in EU countries, where the labor force tends to emigrate (Eurostat, 2020d). However, the Brexit and COVID-19 pandemic and free movement for several Eastern European countries, such as the Ukraine, have offered the Baltic states an opportunity to change their migration dynamics. Estonia's population has shown an increase in population growth of 0.09–0.27 percent since 2015 (World Bank, 2019). 2020 was the first year after a long period when the emigration/immigration ratio was positive in Lithuania and Latvia. Such trends should be supported by ready immigration, training, and integration policies if countries want to capitalize on positive geopolitical trends.

Digital skills. Digitalization-related skills are a broad concept that encompasses various population characteristics believed to be associated with the ability to use digital technologies for value creation. The situation varies heavily in the Baltic states regarding the skills associated with digitalization. Estonia implemented the most successful strategy for developing digital skills. The extent of ICT specialists, measured as a percentage of employment, constitutes 2.7 percent in Lithuania, 4.4 percent in Estonia, and 1.7 percent in Latvia, while the EU average is 3.9 percent (DESI, 2020a). Digital skills among the active population are evaluated at the 28th (Lithuania), 12th (Estonia), and 48th (Latvia) places among one hundred industrial countries (WEF, 2018). Knowledge-intensive employment constitutes 42.3 percent of the working population in the Baltic states (WEF, 2018). The current assessment of digital skills constitutes an opportunity, which has already been seized by Estonia. Education priorities emphasizing digital skills and creativity supported by increased R&D expenditures would help successfully exploit these opportunities for other countries in the Baltic states.

7.10 CONCLUSIONS

Businesses are increasingly adopting digitalization strategies to respond to the disruptive forces of globalization, hypercompetition, and the COVID-19 crisis. Although industry leaders and mega corporations, as early adopters of modern and innovative digital technologies, are enjoying the benefits of Industry 4.0, smaller businesses are struggling to keep up with the digitalization race. Despite the advantages of SMEs in terms of innovation capacity and reactiveness to change, these businesses usually struggle with the complexity and resource intensity of digital transformation. Since SMEs are the backbone of many economies and vital to healthy wealth generation and distribution, European governments are striving to develop and execute supportive digitalization policies to facilitate the digital transformation of SMEs under the Industry 4.0 scenario. Nonetheless, devising an effective supportive digitalization policy relies on a clear understanding of SMEs' needs, strengths, and future strategic priorities for digitalization.

Overall, the performance of the Baltic states in the digitalization of businesses and public services looks promising. The promotion of the digital transformation of SMEs has been an indispensable part of digitalization policies in the Baltic states. As a result, SMEs active in this region have been able to enjoy a well-developed digital infrastructure that can facilitate the adoption of Industry 4.0 digital technologies. SMEs in the Baltic states have the opportunity to embark on the digitalization journey to enjoy higher exposure to the regional and global value chains and react to the uncertainties and complications caused by the COVID-19 pandemic. Nonetheless, barriers such as the lack of necessary human capital, R&D investment, and limited regional business opportunities, have slowed the digitalization pace of SMEs in the Baltic states.

In the case of Lithuanian manufacturing SMEs, the results show mixed situation. The findings reveal that the extent of service innovation and front-end digitalization in SMEs is similar to that in large companies and that SMEs lag in terms of back-end digitalization compared to large companies. Digitalization has been mostly limited to using standard digital technologies for production planning, monitoring, and control for these businesses. Conversely, the adoption and implication of more Industry-4.0-related technologies, such as additive manufacturing, simulation, and industrial robotics, have been significantly limited among Lithuanian manufacturing SMEs. Overall, the Baltic states should develop and implement more progressive and focused SME digital competence development programs to increase the future digitally enabled innovativeness capacity of smaller businesses throughout the ongoing digitalization race.

7.11 ACKNOWLEDGEMENTS

This research has received funding from the European Union's Horizon 2020 research and innovation program under grant agreement No 810318.

REFERENCES

Breznitz, D., Forman, C., & Wen, W. 2018. The role of venture capital in the formation of a new technological ecosystem: Evidence from the cloud. *MIS Quarterly*, 42(4), 1143–1169.

Brill, T. M., Munoz, L., & Miller, R. J. 2019. Siri, Alexa, and other digital assistants: A study of customer satisfaction with artificial intelligence applications. *Journal of Marketing Management*, 35(15–16), 1401–1436.

Brynjolfsson, E., & McAfee, A. 2014. *The second machine age: Work, progress, and prosperity in a time of brilliant technologies*. New York. London. Norton Publishers.

Cheng, M. K. C., Rehman, S., Seneviratne, M. D., & Zhang, S. 2015. *Reaping the benefits from global value chains (No. 15–204)*. Washington, D.C: International Monetary Fund.

Coreynen, W., Matthyssens, P., & Van Bockhaven, W. 2017. Boosting servitization through digitization: Pathways and dynamic resource configurations for manufacturers. *Industrial Marketing Management*, 60, 42–53.

d'Aveni, R. A. 2010. *Beating the commodity trap: How to maximize your competitive position and increase your pricing power*. Boston: Harvard Business Press.

DESI. 2020a. Retrieved https://ec.europa.eu/digital-single-market/en/digital-economy-and-society-index-desi.

DESI. 2020b. Retrieved https://ec.europa.eu/newsroom/dae/document.cfm?doc_id=66911.

DESI. 2020c. Retrieved https://ec.europa.eu/newsroom/dae/document.cfm?doc_id=66922.

DESI. 2020d. Retrieved https://ec.europa.eu/newsroom/dae/document.cfm?doc_id=66919.

Ding, H., Gao, R. X., Isaksson, A. J., Landers, R. G., Parisini, T., & Yuan, Y. 2020. State of AI-based monitoring in smart manufacturing and introduction to focused section. *IEEE/ASME Transactions on Mechatronics*, 25(5), 2143–2154.

Eggert, A., Hogreve, J., Ulaga, W., & Muenkhoff, E. 2011. Industrial services, product innovations, and firm profitability: A multiple-group latent growth curve analysis. *Industrial Marketing Management*, 40(5), 661–670.

Eurostat. 2020a. Real GDP per capita. Retrieved https://ec.europa.eu/eurostat/databrowser/view/sdg_08_10/default/table?lang=en.

Eurostat. 2020b. Nominal labour productivity per person employed. Retrieved https://ec.europa.eu/eurostat/databrowser/view/tec00116/default/table?lang=en.

Eurostat. 2020c. Manufacturing statistics. Retrieved https://appsso.eurostat.ec.europa.eu/nui/submitViewTableAction.do.

Eurostat. 2020d. Wages and labour costs. Retrieved https://ec.europa.eu/eurostat/statistics-explained/index.php/Wages_and_labour_costs.

Evangelista, R., Guerrieri, P., & Meliciani, V. 2014. The economic impact of digital technologies in Europe. *Economics of Innovation and New Technology*, 23(8), 802–824.

Ford, M. 2015. *The rise of the robots: Technology and the threat of mass unemployment*. New York. Basic Books.

Ghobakhloo, M. 2020. Industry 4.0, digitization, and opportunities for sustainability. *Journal of Cleaner Production*, 252, 119869.

Hasselbring, W. 2000. Information system integration. *Communications of the ACM*, 43(6), 32–38.

Horvath, D., & Szabo, R. Z. 2019. Driving forces and barriers of Industry 4.0: Do multinational and small and medium-sized companies have equal opportunities? *Technological Forecasting and Social Change*, 146, 119–132.

Iansiti, M., & Lakhani, K. M. 2014. Digital ubiquity: How connections, sensors, and data are revolutionizing business. *Harvard Business Review*, 92(11), 90–99.

Law, C. C., & Ngai, E. W. 2007. ERP systems adoption: An exploratory study of the organizational factors and impacts of ERP success. *Information & Management*, 44(4), 418–432.

Lenka, S., Parida, V., & Wincent, J. 2017. Digitalization capabilities as enablers of value co-creation in servitizing firms. *Psychology & Marketing*, 34(1), 92–100.

Lightfoot, H., Baines, T., & Smart, P. 2013. The servitization of manufacturing: A systematic literature review of interdependent trends. *International Journal of Operations & Production Management*, 33(11/12), 1408–1434.

LSD, Department of Statistics of Lithuania. 2019. Retrieved https://osp.stat.gov.lt/lietuvos-statistikos-metrastis/lsm-2019/verslas/pramone.

Maier II, J. B., & Walker, D. A. 1987. The role of venture capital in financing small business. *Journal of Business Venturing*, 2(3), 207–214.

Manyika, J., Sinclair, J., Dobbs, R., Strube, G., Rassey, L., Mischke, J., & Ramaswamy, S. 2012. *Manufacturing the future: the next era of global growth and innovation*. London: McKinsey Global Institute.

Mittal, S., Khan, M. A., Purohit, J. K., Menon, K., Romero, D., & Wuest, T. 2020. A smart manufacturing adoption framework for SMEs. *International Journal of Production Research*, 58(5), 1555–1573.

Moeuf, A., Lamouri, S., Pellerin, R., Tamayo-Giraldo, S., Tobon-Valencia, E., & Eburdy, R. 2020. Identification of critical success factors, risks and opportunities of industry 4.0 in SMEs. *International Journal of Production Research*, 58(5), 1384–1400.

Müller, J. M., Buliga, O., & Voigt, K.-I. 2018. Fortune favors the prepared: How SMEs approach business model innovations in industry 4.0. *Technological Forecasting and Social Change*, 132, 2–17.

Müller, J. M., Buliga, O., & Voigt, K. I. 2020a. The role of absorptive capacity and innovation strategy in the design of industry 4.0 business models-A comparison between SMEs and large enterprises. *European Management Journal*. https://doi.org/10.1016/j.emj.2020.01.002.

Müller, J. M., Veile, J. W., & Voigt, K. I. 2020b. Prerequisites and incentives for digital information sharing in Industry 4.0 – An international comparison across data types. *Computers & Industrial Engineering*, 148, 106733.

Observatory of Economic Complexity. 2020. Retrieved https://oec.world/en/profile/country/ltu.

OECD. 2020a. Gross domestic spending on R&D. Retrieved https://data.oecd.org/rd/gross-domestic-spending-on-r-d.htm.

Raj, A., Dwivedi, G., Sharma, A., Lopes de Sousa Jabbour, A. B., & Rajak, S. 2020. Barriers to the adoption of industry 4.0 technologies in the manufacturing sector: An inter-country comparative perspective. *International Journal of Production Economics*, 224, 107546.

Roland Berger. 2014. Industry 4.0. The new industrial revolution. How Europe will succeed. Roland Berger strategy consultants.

Sahi, G. K., Gupta, M. C., & Cheng, T. C. E. 2020. The effects of strategic orientation on operational ambidexterity: A study of Indian SMEs in the industry 4.0 era. *International Journal of Production Economics*. doi:10.1016/j.ijpe.2019.05.014.

Salvador, F., De Holan, P. M., & Piller, F. 2009. Cracking the code of mass customization. *MIT Sloan Management Review*, 50(3), 71–78.

Schlaepfer, R. C., Koch, M., & Merkhofer, P. 2015. Industry 4.0 challenges and solutions for the digital transformation and use of exponential technologies. Deloitte Report.
Siggelkow, N., & Terwiesch, C. 2019. *Connected strategy: Building continuous customer relationships for competitive advantage*. Boston: Harvard Business Press.
Statista. 2020. Population of Estonia, Latvia and Lithuania from 1950 to 2020. Retrieved www.statista.com/statistics/1016444/total-population-baltic-states-1950-2020/.
Ulaga, W., & Reinartz, W. J. 2011. Hybrid offerings: How manufacturing firms combine goods and services successfully. *Journal of Marketing*, 75(6), 5–23.
Vilkas, M., Rauleckas, R., Šeinauskienė, B., & Rutelionė, A. 2021. Lean, agile and service-oriented performers: Templates of organising in a global production field. *Total Quality Management & Business Excellence*, 32(9–10), 1122–1146.
Wang, S., Cavusoglu, H., & Deng, Z. 2016. Early mover advantage in e-commerce platforms with low entry barriers: The role of customer relationship management capabilities. *Information & Management*, 53(2), 197–206.
WEF. 2018. *The readiness for the future of production report 2018*. Cologny/Geneva: World Economic Forum.
World Bank. 2019. Population growth. Retrieved https://data.worldbank.org/indicator/SP.POP.GROW.

8 Industry 4.0 in Poland from a Perspective of SMEs

Anna Michna and Jan Kaźmierczak

CONTENTS

8.1 INTRODUCTION

Every revolution in human history, no matter whether social, religious, or connected with changes in the area of human creative activity, or the broadly understood art or strictly utilitarian activity, can and should be considered in many dimensions. Every period of revolutionary changes in any of these areas (such transformations could take place simultaneously and were driven by one another) had its most general dimension (we would call it "global" nowadays), but it also covered transformations in particular parts of the known world (i.e., continents and countries now), as well as in larger and smaller communities/social groups, also shaping the fate of individuals embedded in the society in one sense or another.

Even if we focus the considerations initiated in the period starting from the second half of 18th century, when the process of technology, economic, social and cultural changes, termed Industrial Revolution, started in England and Scotland, we will relatively easily see the similarities but also differences of that process's impact on societies and individuals in different parts of what at that time was the center of the world, i.e., the European continent. Those differences were determined by different political regimes, the structure and the education level of the society, as well as the resources held and the orientation of economies in particular countries of Europe in those days connected with those resources.

DOI: 10.1201/9781003165880-8

We also live in the times of great (revolutionary?) transformations that shape the economy of the world and its specific parts. We are speaking of the so-called fourth Industrial Revolution, usually termed Industry 4.0 (I4.0). This revolution uses technologies of collecting, processing, sending, and using data/information. Published works concerning the said revolution discuss both its technical (Braun *et al.* 2018) and other aspects (Cellary 2019; Grewiński 2018). In the reference works, we can find, for instance, lists of technologies that change industrial production (Erboz 2017). An interesting "division" of the I4.0 paradigm is proposed also by the article authors (Weyer *et al.* 2015).

Some reference works also contain many studies describing the specific relations, connecting the I4.0 concept with global economy sectors or economies of individual countries worldwide with all specific conditions stemming from the local tradition, culture, resources and, last but not least, all human behaviors that may imply using Industry 4.0 (Grzelczak *et al.* 2017a), (Kosacka-Olejnik and Pitakaso 2019).

Many authors have already published works depicting, for instance, the results of comparing the conditions for implementing I4.0 in Poland in reference to other countries (Grzyb 2019), (Piątkowski 2020) as well as to internal conditions (Grzelczak *et al.* 2017b). In what follows, you can find a subjective perspective of this study's authors concerning the factors determining the implementation of I4.0 concept in the Polish society and economy.

8.2 POLAND: A GENERAL BACKGROUND

Poland is a medium-sized country (with the area of 312,696 km², population of 37.97 million (2019)). Poland has a three-tier administrative division. It is divided into 16 regions (voivodeships) with poviat districts (314 poviat districts and 66 cities with poviat rights which are communes but also perform the tasks of poviat districts). All poviat districts are composed of 2,477 communes (as of 1 January 2020). The administrative map of Poland is presented in Figure 8.1.

For many reasons, resulting from our more or less recent history, Poland is a relatively highly differentiated country also in terms of economy. In the 16th, 17th, and 18th centuries, Poland was called a "breadbasket of Europe." Its economy and the social and political system were strongly shaped by agriculture. In our modern history, after World War II, Poland belonged to the zone governed by the former Soviet Union, and its economy was dominated by the Socialist doctrine of nationalized economy and central planning. A factor that differentiated Poland from other "real socialist countries" of those days was a significant share of small (and no larger than medium-sized) private companies, including but not limited to family-owned ones, in the economic life. This is why Poland coped better in the market economy conditions, especially in the SME sector, in the period of transformations in 1989–1990. On the other hand, large state-owned companies underwent the long transformation process, difficult in economic and social terms, which resulted in turning a significant part of the Socialist economy into a strong industrial sector operating based on modern principles.

At present, large companies in Poland, some with foreign capital, are fertile soil for the practical implementation of new concepts and solutions. This includes the

FIGURE 8.1 Administrative map of Poland.

Source: gt29/Shutterstock.com

implementation of the Industry 4.0 paradigm. According to the authors of this study, large Polish companies do not deviate significantly from the ones operating in other European economies. This is proved by the development strategies of companies such as Kombinat Górniczo – Hutniczy Miedzi SA (Wirth *et al.* 2010).

The fertility and openness to the new concepts and technologies are slightly different in the sector of Polish SMEs (Kaźmierczak 2019). What is clear here is the entrepreneurs' fears connected with the need to adapt the company to the operation in conditions determined by the I4.0 paradigm on the one hand, and the relatively widespread belief of owners and managers of small and medium-sized companies that the processes connected with the fourth Industrial Revolution do not refer to them, on the other. An undoubtedly significant challenge is faced primarily by academic centers that should offer adequate measures and ways of complementing knowledge and expert support to entrepreneurs when implementing the undertaken tasks.

It should also be stressed here that a prerequisite for the successful adaptation to the requirements of I4.0 also involves a number of challenges connected with a broadly understood business environment. Digitization of this environment is a major challenge of that type.

8.3 POLAND: PROCESSES OF DIGITIZATION POLICY

In the 1990s, the digitization and computerization processes in different areas of social functioning were highly intensified in Poland. This chance was used to initiate broad discussions on the digitization impact on social behaviors and functioning of society. Digitization-related activities referred primarily to such areas as the legislation and legal sector (Karolczyk 2018), field information management (Drzewiecki 2008), as well as the broadly understood central and local administration (Śliwczyński 2015). Research works emerged, and, consequently, so did practical activities connected with cybersecurity (Gryszczyńska 2016). Polish researchers also dealt with the digitization impact on the political parties' functioning (Jacuński 2019). Studies and practical measures emerged also in the education system (Mazurkiewicz 2020).

In summary, the process of digitization has appeared and is continued both in Polish public and private sector. However, the progress and development of this process significantly differs between sectors of national economy. And, importantly, this process is much more intensive and effective in large industrial companies operating in Poland than in Polish SMEs. For some time, researchers have tried to identify the nature of particular problems with implementing I4.0 in the SME sector – their origins as well as potential results of the identified problems. The empirical studies of such problems have been and are being carried out not only in Poland (Kaźmierczak 2019) but also in other European countries (Sommer 2015).

8.4 METHODOLOGY OF EMPIRICAL RESEARCH

The empirical studies aimed at identifying obstacles to implementing Industry 4.0 solutions and learning the plans concerning the implementation of those solutions in the Polish SMEs carrying out manufacturing activities.

The empirical studies covered a group of small and medium-sized manufacturing enterprises from different parts of Poland. Using a proprietary survey addressed to the SME managers, data was collected for 562 enterprises (the study took place from December 2019 to March 2020). The participants provided anonymous responses to the questions using a five-point Likert scale, where: 1=strongly disagree, 2=disagree, 3=neutral, 4=agree and 5=strongly agree. Because of the number of employees, the study sample comprised 265 microenterprises (employing up to 10 employees), 221 small enterprises (10–49 employees) and 76 medium-sized companies (50–249 employees). When it came to the length of the company's operations, the study sample comprised enterprises existing for 5 years (167 entities), from 6 to 10 years (79 entities), from 11 to 15 years (75 entities), from 16 to 20 years (61 entities), and more than 20 years (180 entities). The core activities based on the PKD (classification of business activities in Poland) mentioned by most companies included repair, maintenance, and installation of machines and equipment (131), manufacture of other goods (98), manufacture of machines and equipment not otherwise classified (49), and the manufacture of finished metal goods, excluding machines and equipment (46).

8.5 INDUSTRY 4.0 BARRIERS

The analysis of obstacles to the Industry 4.0 implementation in SMEs included the following:

- The following are considered main obstacles to the implementation of Industry 4.0 solutions (including robotics and production automation) in our company:
  - The need to employ specialists (B1a).
  - Reluctance to changes (B1b).
  - The need to dismiss some employees (B1c).
  - The need to improve qualifications of the existing employees (B1d).
  - The investment risk (B1e).
- The major obstacles to the implementation of cutting-edge technology in the customer service area include the limitations:
  - Limitations on the company's part (e.g., the absence of funds, low employee competences, no expertise) (B2a).
  - Limitations on the customers' part (e.g., reluctance to use new technology, no expertise, no trust to mobile devices and artificial intelligence) (B2b).

Descriptive statistics and the correlation matrix for items B1a to B2b are presented in Table 8.1.

TABLE 8.1
Descriptive Statistics and Correlation Matrix for Items B1a–B2d

	Mean	SD	Median	B1a	B1b	B1c	B1d	B1d	B2a	B2b
All SMEs										
B1a	2.97	1.44	3.00	1.00						
B1b	2.36	1.29	2.00	0.29*	1.00					
B1c	1.98	1.11	2.00	0.28*	0.49*	1.00				
B1d	2.82	1.38	3.00	0.38*	0.35*	0.37*	1.00			
B1e	3.32	1.33	4.00	0.33*	0.22*	0.22*	0.33*	1.00		
B2a	3.38	1.33	4.00	0.26*	0.12*	0.08	0.12*	0.31*	1.00	
B2b	3.10	1.17	3.00	0.17*	0.18*	0.14*	0.13*	0.12*	0.20*	1.00

(Continued)

TABLE 8.1 (continued)
Descriptive Statistics and Correlation Matrix for Items B1a–B2d

	Mean	SD	Median	B1a	B1b	B1c	B1d	B1d	B2a	B2b
Microenterprises										
B1a	2.72	1.49	3.00	1.00						
B1b	2.21	1.24	2.00	0.21*	1.00					
B1c	1.83	1.08	1.00	0.24*	0.49*	1.00				
B1d	2.57	1.38	3.00	0.40*	0.35*	0.35*	1.00			
B1e	3.17	1.43	3.00	0.43*	0.24*	0.20*	0.31*	1.00		
B2a	3.41	1.33	4.00	0.32*	0.06	0.10	0.12	0.34*	1.00	
B2b	3.05	1.15	3.00	0.15*	0.11	0.06	0.12	0.13*	0.13*	1.00
Small enterprises										
B1a	3.21	1.37	4.00	1.00						
B1b	2.46	1.31	2.00	0.37*	1.00					
B1c	2.13	1.17	2.00	0.31*	0.46*	1.00				
B1d	3.10	1.37	3.00	0.35*	0.30*	0.39*	1.00			
B1e	3.46	1.20	4.00	0.23*	0.21*	0.23*	0.39*	1.00		
B2a	3.35	1.37	4.00	0.20*	0.15*	0.06	0.15*	0.31*	1.00	
B2b	3.14	1.22	3.00	0.24*	0.29*	0.25*	0.20*	0.18*	0.30*	1.00
Medium enterprises										
B1a	3.14	1.36	3.00	1.00						
B1b	2.57	1.32	2.00	0.25*	1.00					
B1c	2.07	1.00	2.00	0.23*	0.53*	1.00				
B1d	2.89	1.30	3.00	0.23*	0.42*	0.25*	1.00			
B1e	3.43	1.27	3.50	0.08	0.10	0.19	0.19	1.00		
B2a	3.34	1.28	4.00	0.24*	0.27*	0.11	0.11	0.19	1.00	
B2b	3.18	1.07	3.00	–0.05	0.04	0.00	–0.08	–0.15	0.16	1.00

Note: *Correlation significant at $p < 0.05$

Regardless of the enterprise size, the correlation between items B1b and B1c was the strongest. This means that in the context of obstacles to implementing Industry 4.0 solutions in a company, the reluctance to changes is frequently accompanied by the need to dismiss some workers.

The highest mean level in the group of all SMEs and microenterprises was reached by B2a obstacle which meant that the limitations on the company's part are the most important among the analyzed obstacles according to respondents, including the absence of funds, low employee competences and the absence of expertise. For small and medium-sized enterprises, the most important obstacle to implementing Industry 4.0 is the investment risk (B1e). The least important obstacle in all groups of companies is the need to dismiss some employees (B1c).

The Kruskal-Wallis test by ranks, median test and the post-hoc test were used to check if there are any statistically important differences between the groups of companies. The tests indicated that there are significant differences for questions Ba1 and B1d. The results prove that when compared to microenterprises, obstacles connected with the need to employ specialists (B1a) and improve qualifications of the existing employees (B1d) are more important in small enterprises.

8.6 INDUSTRY 4.0 DRIVERS

When analyzing the factors stimulating the implementation of Industry 4.0 in SMEs (drivers), the following prompts from the questionnaire were taken into account:

- The enterprise has a research and development unit (D1) in its structure.
- Competition in the industry, in which the organisation operates, is stronger than in previous years (D2).
- Our enterprise intensively cooperates in its innovative activities with:
    - Customers and buyers (D3a).
    - Suppliers (D3b).
    - R&D units, institutes, and universities (D3c).
    - Competitors (D3d).

Descriptive statistics and the correlation matrix for items D1 to D3d are presented in Table 8.2. Statistically significant correlation coefficients at the significance level of $p < 0.05$ are marked in the table with an asterisk (*).

TABLE 8.2
Descriptive Statistics and a Correlation Matrix for Items D1–D3d

	Mean	SD	Median	D1	D2	D3a	D3b	D3c	D3d
All SMEs									
D1	1.96	1.48	1.00	1.00					
D2	3.69	1.22	4.00	0.10*	1.00				
D3a	3.81	1.36	4.00	0.21*	0.16*	1.00			
D3b	3.46	1.42	4.00	0.14*	0.15*	0.52*	1.00		
D3c	2.21	1.43	2.00	0.35*	0.03	0.21*	0.19*	1.00	
D3d	2.06	1.24	2.00	0.12*	0.09*	0.28*	0.26*	0.27*	1.00
Microenterprises									
D1	1.65	1.28	1.00	1.00					
D2	3.64	1.25	4.00	0.03	1.00				
D3a	3.77	1.44	4.00	0.14*	0.08	1.00			
D3b	3.32	1.45	4.00	0.11	0.16*	0.51*	1.00		
D3c	1.91	1.29	1.00	0.25*	-0.05	0.15*	0.12	1.00	
D3d	1.99	1.23	1.00	0.10	0.07	0.31*	0.21*	0.28*	1.00

(*Continued*)

TABLE 8.2 (continued)
Descriptive Statistics and a Correlation Matrix for Items D1–D3d

	Mean	SD	Median	D1	D2	D3a	D3b	D3c	D3d
Small enterprises									
D1	2.13	1.54	1.00	1.00					
D2	3.81	1.13	4.00	0.17*	1.00				
D3a	3.90	1.22	4.00	0.25*	0.22*	1.00			
D3b	3.63	1.35	4.00	0.15*	0.15*	0.48*	1.00		
D3c	2.32	1.43	2.00	0.35*	0.11	0.21*	0.24*	1.00	
D3d	2.21	1.28	2.00	0.23*	0.08	0.27*	0.36*	0.33*	1.00
Medium enterprises									
D1	2.55	1.73	1.50	1.00					
D2	3.50	1.35	4.00	0.16	1.00				
D3a	3.72	1.45	4.00	0.35*	0.24*	1.00			
D3b	3.43	1.44	4.00	0.14	0.09	0.62*	1.00		
D3c	2.91	1.58	3.00	0.37*	0.12	0.40*	0.26*	1.00	
D3d	1.84	1.12	1.00	–0.14	0.11	0.23*	0.15	0.14	1.00

Note: *Correlation significant at $p < 0.05$

The analysis of the correlation matrix shows that, regardless of the size of the enterprise, the correlation between questions D3a and D3b is the strongest. This means that cooperation in the field of innovative activities with customers and buyers is often accompanied by cooperation in this area with suppliers. However, a weak or no correlation was observed between the competition in the industry (question D2) and the remaining analyzed questions.

In all groups of enterprises, the question concerning cooperation in innovative activities with customers and buyers has the highest average value. Next, it was examined whether there were significant statistical differences between these groups. The Kruskal-Wallis test (one-way ANOVA test on ranks), the median test and the Post-hoc test (multiple comparison of mean ranks) were used. The tests showed that there were significant differences for questions D1 and D3c. These results, together with the analysis of the values of descriptive statistics, indicate that:

- Microenterprises less frequently have a research and development unit (D1) in their structure than small and medium-sized enterprises.
- The larger the enterprise, the more intensively it cooperates in its innovative activities with R&D units, institutes, and universities (D3c). In this case, the difference is statistically significant, not only between micro and small enterprises, but also between small and medium-sized enterprises.

8.7 INDUSTRY 4.0 OPPORTUNITIES

Analyzing the opportunities and plans concerning the implementation of Industry 4.0, responses to the following questions were considered:

- Our company intends to increase the expenditure on investments connected with Industry 4.0 in 5 years to come (P1).
- In 3–5 years to come, our company plans:
  - To modernize the machines held significantly (P2a).
  - To increase digitization and computerization significantly (P2b).
  - To increase the automation degree (P2c).
  - To introduce autonomous robots or to increase their number (P2d).
  - To introduce additive manufacturing or to extend it (3D printing) (P2e).
  - To introduce the virtual technology application (VR; virtual reality, AR; augmented reality) or to extend it (P2f).
  - To increase the technical personnel's competences significantly (P2g).

Descriptive statistics and the correlation matrix for items P1 to P2g are presented in Table 8.3.

TABLE 8.3
Descriptive Statistics and Correlation Matrix for Items P1–P2g

	Mean	SD	Median	P1	P2a	P2b	P2c	P2d	P2e	P2f	P2g
All SMEs											
P1	3.14	1.27	3.00	1.00							
P2a	3.12	1.37	3.00	0.46*	1.00						
P2b	3.28	1.27	4.00	0.54*	0.54*	1.00					
P2c	2.97	1.36	3.00	0.50*	0.56*	0.70*	1.00				
P2d	1.92	1.26	1.00	0.43*	0.48*	0.43*	0.57*	1.00			
P2e	1.97	1.28	1.00	0.37*	0.33*	0.39*	0.37*	0.58*	1.00		
P2f	1.96	1.21	1.00	0.32*	0.29*	0.43*	0.40*	0.57*	0.63*	1.00	
P2g	3.41	1.23	4.00	0.44*	0.47*	0.55*	0.45*	0.39*	0.34*	0.38*	1.00
Microenterprises											
P1	2.86	1.31	3.00	1.00							
P2a	2.86	1.42	3.00	0.41*	1.00						
P2b	3.01	1.35	3.00	0.54*	0.56*	1.00					
P2c	2.62	1.39	2.00	0.50*	0.51*	0.68*	1.00				
P2d	1.64	1.13	1.00	0.39*	0.47*	0.42*	0.53*	1.00			
P2e	1.85	1.29	1.00	0.39*	0.36*	0.42*	0.38*	0.66*	1.00		
P2f	1.76	1.15	1.00	0.29*	0.31*	0.47*	0.45*	0.64*	0.64*	1.00	
P2g	3.19	1.33	3.00	0.39*	0.40*	0.59*	0.42*	0.36*	0.33*	0.40*	1.00

(Continued)

TABLE 8.3 (continued)
Descriptive Statistics and Correlation Matrix for Items P1–P2g

	Mean	SD	Median	P1	P2a	P2b	P2c	P2d	P2e	P2f	P2g
Small enterprises											
P1	3.37	1.16	3.00	1.00							
P2a	3.26	1.33	4.00	0.46*	1.00						
P2b	3.50	1.17	4.00	0.51*	0.48*	1.00					
P2c	3.22	1.28	3.00	0.45*	0.55*	0.72*	1.00				
P2d	2.15	1.32	2.00	0.42*	0.47*	0.40*	0.60*	1.00			
P2e	2.10	1.26	2.00	0.34*	0.30*	0.32*	0.38*	0.58*	1.00		
P2f	2.15	1.22	2.00	0.29*	0.26*	0.36*	0.35*	0.52*	0.61*	1.00	
P2g	3.56	1.16	4.00	0.53*	0.56*	0.52*	0.46*	0.38*	0.37*	0.37*	1.00
Medium enterprises											
P1	3.45	1.24	4.00	1.00							
P2a	3.62	1.08	4.00	0.49*	1.00						
P2b	3.55	1.10	4.00	0.47*	0.54*	1.00					
P2c	3.43	1.23	4.00	0.43*	0.67*	0.61*	1.00				
P2d	2.24	1.28	2.00	0.40*	0.42*	0.38*	0.48*	1.00			
P2e	1.99	1.28	1.00	0.37*	0.26*	0.37*	0.20	0.37*	1.00		
P2f	2.07	1.31	1.50	0.38*	0.18	0.35*	0.22	0.46*	0.64*	1.00	
P2g	3.76	0.89	4.00	0.25*	0.37*	0.19	0.28*	0.36*	0.23*	0.28*	1.00

Note: *Correlation significant at $p < 0.05$

The correlation coefficient is the highest among all the SMEs analyzed and in the group of micro- and small companies for the correlation between questions P2b and P2c. This means that the plans concerning the increase in the digitization and computerization (P2b) often overlap with the plans to increase the automation degree

(P2c) in the company. The plans to modernize the machines held (P2a) and the plans to introduce or extend virtual technology (P2f) display the weakest correlation.

The highest mean value in all groups of companies was reached by P2g question, meaning the enterprises attach the greatest importance to the plans of increasing the technical personnel's competences. The smallest interest among all the SMEs and microenterprises is enjoyed by the plans to introduce autonomous robots or to increase their number (P2d).

The question of whether there were any significant differences in the plans concerning Industry 4.0 in the groups of companies of differing sizes was studied. The Kruskal-Wallis test revealed there are differences between groups of companies for all questions except for P2b and P2g. The results of the Post-hoc test (multiple comparison of mean ranks) and descriptive statistics indicate that microenterprises have less advanced plans to implement Industry 4.0 solutions when compared to small ones (this refers to questions P2e and P2f) and when compared both to the small and medium-sized enterprises (this refers to questions P1, P2a, P2c and P2d).

TABLE 8.4
A Correlation Matrix for Items D1–D3d and P1–P2g

	P1	P2a	P2b	P2c	P2d	P2e	P2f	P2g
D1	0.24*	0.25*	0.24*	0.22*	0.29*	0.30*	0.23*	0.23*
D2	0.13*	0.10*	0.15*	0.10*	0.10*	0.17*	0.16*	0.19*
D3a	0.30*	0.16*	0.22*	0.20*	0.15*	0.17*	0.15*	0.30*
D3b	0.28*	0.22*	0.24*	0.22*	0.19*	0.18*	0.18*	0.24*
D3c	0.29*	0.27*	0.27*	0.25*	0.34*	0.27*	0.31*	0.27*
D3d	0.15*	0.10*	0.18*	0.15*	0.20*	0.17*	0.25*	0.18*

Note: *Correlation significant at $p < 0.05$

On the basis of the correlation matrix, the relationships between the factors instrumental to the implementation of Industry 4.0 solutions and the plans for implementing Industry 4.0 solutions were examined (Table 8.4). All the correlation coefficients in Table 8.4 are statistically significant, but the strongest correlation (0.34) exists between the enterprise's cooperation with R&D units, institutes and universities (D3c) and plans to introduce or increase the number of autonomous robots (P2d).

8.8 CONCLUSIONS

In the ranking by Digital Economy and Society Index (DESI), Poland took 23rd place in the group of 28 EU member states in 2020 (European Commisson 2020), ahead of Italy, Bulgaria, Greece, and Romania. When compared to previous years, the position of Poland in the said ranking improved (in 2019 it was 25th, and in 2018 it took 24th position). Unfortunately, the basic digital skills remain low when compared to the EU average. They must be improved even more as they are required in the times of COVID-19. Poland has a good level of using mobile broadband services and competitive prices in this respect. However, the degree of using digital technology proves that as much as 60 percent of companies are characterized by a very low digitization level (EU average is 39 percent) and only 11 percent of them are companies with high digitization level (EU: 26 percent).

Based on the empirical studies carried out, a conclusion can be drawn that the most frequent obstacle to implementing Industry 4.0 solutions by SMEs is the perception of investments in Industry 4.0 as risk laden. Another important obstacle is the need to employ specialists. Those obstacles are closely correlated for microenterprises. The limitations related to implementing cutting-edge technology in the customer service area include the ones on the company's part (e.g., absence of funds, low employee competences, absence of expertise). With respect to the investment plan scope, the highest priority is given to the significant increase in the technical personnel's competencies, particularly among medium-sized companies.

Certain limitations of the presented empirical study results stem, firstly, from allocating results to the selected research sample, which, though relatively numerous, does not ensure the results would be identical for another group. Secondly, it stems from the timeframe, as the studies were carried out right before the outbreak of COVID-19 epidemic and the resulting lockdown in Poland. To offer a broader picture of the SME situation in the context of Industry 4.0, the studies should be repeated, since the pandemic influenced the business activity conditions significantly and many organizations, even if they were not prepared for it, were forced to initiate digital activities in this unprecedented situation. Applying Industry 4.0 solutions also contributed to improved health-related safety and minimized economic effects of the pandemics and will help to recover from the crisis faster. Moreover, it should be claimed that today's experience, regardless of future epidemic-related situations, will contribute to additional lasting changes in organizations' operation.

The directions of further studies devoted to the implementation of Industry 4.0 in Polish SMEs running manufacturing activity could consider the role of cooperation between organizations (Michna *et al.* 2020), including both the sector-specific ones (e.g., with competitors) and other (e.g., with universities). Such cooperation might help to minimize the limited resources of SMEs (Kmieciak and Michna 2018). What is more, it would be interesting to see if sharing expertise between the organization participants (Michna *et al.* 2018) contributes to implementing Industry 4.0 solutions.

REFERENCES

Braun, A., Ohlhausen, P., Alt, C., Bahlinger, D., Chaves, D. C., Egeler, M., . . . Weber, C. (2018). The way to the Industry 4.0 roadmap. *ZWF Zeitschrift Fuer Wirtschaftlichen Fabrikbetrieb*, 113(4), 254–257. Available from: https://doi.org/10.3139/104.111888 [Accessed 19 December 2020].

Cellary, W. (2019). Non-technical challenges of industry 4.0. *IFIP Advances in Information and Communication Technology*, 568, 3–10. Springer New York LLC. Available from: https://doi.org/10.1007/978-3-030-28464-0_1 [Accessed 19 December 2020].

Drzewiecki, W. (2008). Sustainable land-use planning support by gis-based evaluation of landscape functions and potentials. *International Archives of the Photogrammetry, Remote Sensing and Spatial Information Sciences – ISPRS Archives*, 37, 1497–1502.

Erboz, G. (2017). "How to define industry 4.0: main pillars of industry 4.0." *International Scientifis Conference Managerial Trends in the Development of Enterprises in Globalization Era*, eds. Košičiarová, I. & Kádeková, Z. Nitra: Slovak University of Agriculture in Nitra, Faculty of Economics and Management, 761–767.

European Commisson (2020). The Digital Economy and Society Index (DESI). Available from: https://ec.europa.eu/digital-single-market/en/desi [Accessed 19 December 2020].

Grewiński, M. (2018). Cyfryzacja i innowacje społeczne – perspektywy i zagrożenia dla społeczeństwa. *Kwartalnik Nauk o Przedsiębiorstwie*, 46(1), 19–29. Available from: https://doi.org/10.5604/01.3001.0012.0980 [Accessed 19 December 2020].

Gryszczyńska, A. (2016). "Cybersecurity of public registers in Poland: Selected legal issues." *Geographic Information Systems Conference and Exhibition "Gis Odyssey 2016,"* eds. Bieda, A., Bydłosz, J., & Kowalczyk, A. Zagreb: Croatian Information Technology Society – GIS Forum, 105–113.

Grzelczak, A., Kosacka, M., & Werner-Lewandowska, K. (2017a). Employees competences for industry 4.0 in Poland – preliminary research results. *24th International Conference on Production Research, ICPR 2017. DEStech Publications*, 139–144.

Grzelczak, A., Werner-Lewandowska, K., & Kosacka, M. (2017b). Perspectives of industry 4.0 development in Poland – Preliminary research results. *24th International Conference on Production Research ICPR 2017, DEStech Publications*. Available from: https://doi.org/10.12783/dtetr/icpr2017/17597 [Accessed 19 December 2020].

Grzyb, K. (2019). Industry 4.0 Market in Poland from the international perspective. *Proceedings of the International Scientific Conference Hradec Economic Days*, part I., 9, 252–262. University of Hradec Kralove. Available from: https://doi.org/10.36689/uhk/hed/2019-01-025 [Accessed 19 December 2020].

Jacuński, M. (2019). Digitalization and Political party life in Poland – a study of selected communication habits of party members and elective representatives. *Polish Political Science Review*, 6(2), 6–25. Available from: https://doi.org/10.2478/ppsr-2018-0011 [Accessed 19 December 2020].

Karolczyk, B. (2018). Informatization of the civil justice system in Poland: An overview of recent changes. *Transformation of Civil Justice*, ed. Uzelac, A., & van Rhee, C. H., 99–117. Available from: https://doi.org/10.1007/978-3-319-97358-6_6 [Accessed 19 December 2020].

Kaźmierczak, J. (2019). SME's versus "Industry 4.0": Extended case study based on results of initial research in Polish conditions. *Multidisciplinary Aspects of Production Engineering*, 2(1), 305–314. Available from: https://doi.org/10.2478/mape-2019-0030 [Accessed 19 December 2020].

Kmieciak, R., & Michna, A. (2018). Knowledge management orientation, innovativeness, and competitive intensity: Evidence from Polish SMEs. *Knowledge Management Research & Practice*, 16(4), 559–572. Available from: https://doi.org/10.1080/14778238.2018.1514997 [Accessed 19 December 2020].

Kosacka-Olejnik, M., & Pitakaso, R. (2019). Industry 4.0: State of the art and research implications. Przemysł 4.0: Stan obecny i wytyczne w zakresie potencjalnych badań. *Logforum*, 15(4), 475–485. Available from: https://doi.org/10.17270/J.LOG.2019.363 [Accessed 19 December 2020].

Mazurkiewicz, A. (2020). Cyfryzacja szkoły w społeczeństwie informacyjnym (na przykładzie polskich rządowych dokumentów i projektów). *Acta Universitatis Lodziensis. Folia Litteraria Polonica*, 56(1), 101–126. Available from: https://doi.org/10.18778/1505-9057.56.07 [Accessed 19 December 2020].

Michna, A., Kmieciak, R., & Brzostek, K. (2018). Development of knowledge management processes in a small organization: A case study. *12th annual International Technology, Education and Development Conference*. Available from: https://doi.org/10.21125/inted.2018.0596 [Accessed 8 December 2020].

Michna, A., Kmieciak, R., & Czerwińska-Lubszczyk, A. (2020). Dimensions of intercompany cooperation in the construction industry and their relations to performance of SMEs. *Engineering Economics*, 31(2), 221–232. Available from: http://dx.doi.org/10.5755/j01.ee.31.2.21212 [Accessed 8 December 2020].

Piątkowski, M. J. (2020). Expectations and challenges in the labour market in the context of industrial revolution 4.0. the agglomeration method-based analysis for Poland and Other EU Member States. *Sustainability*, 12(13). Available from: https://doi.org/10.3390/su12135437 [Accessed 19 December 2020].

Sommer, L. (2015). Industrial revolution – industry 4.0: Are German manufacturing SMEs the first victims of this revolution? *Journal of Industrial Engineering and Management*, 8(5), 1512–1532. Available from: https://doi.org/10.3926/jiem.1470 [Accessed 19 December 2020].

Śliwczyński, B. (2015). E-invoicing system in Poland. *E-Mentor*, 4(61), 75–83. Available from: https://doi.org/10.15219/em61.1200 [Accessed 19 December 2020].

Weyer, S., Schmitt, M., Ohmer, M., & Gorecky, D. (2015). Towards industry 4.0 – Standardization as the crucial challenge for highly modular, multi-vendor production systems. *In IFAC-PapersOnLine*, 28, 579–584. Available from: https://doi.org/10.1016/j.ifacol.2015.06.143 [Accessed 19 December 2020].

Wirth, H., Kubacki, K., & Ziemkiewicz, J. (2010). Strategia KGHM Polska Miedź SA na lata 2009–2018. *Górnictwo i Geologia*, 5, 169–179.

9 Industry 4.0

Present or Future in Serbia

Zoran Vitorovich

CONTENTS

9.1 INTRODUCTION

Never in human history has a technological breakthrough been recorded as in the last few years, and these changes have far-reaching consequences in the way civilization will develop. The impact of the globalization process on society, economics, politics, and climate change is great, and technological advances have been unprecedented. Technological development has imposed an additional dimension on comprehensive informatization in the form of a new industrial revolution, known as Industry 4.0.

Although the term is slowly running through the Serbian media, a good part of the developed world is in what is called the fourth Industrial Revolution. The academic community is trying to generate awareness of the need to transform the current education system and industry.

The aim of this chapter is to get closer to the general public what Industry 4.0 is, on what principles it is based, what it means for Serbian industry and the economy as a whole, what it means for humans, existing, and future occupations, where the developed world is currently and where it aspires to be.

The subject of the research work is to determine the position of Serbian SME CA compared to 4.0 Industry and members of the European Community.

9.2 COUNTRY BACKGROUND

According to the last constitution from 2006, the Republic of Serbia is a state based on the rule of law and social justice, the principles of civil democracy, human and

DOI: 10.1201/9781003165880-9

minority rights and freedoms, and belonging to European principles and values (Article 1). Serbia is a unitary state, with a multiparty system and direct elections (Article 2). According to the constitution, respect for human and civil rights and freedoms is guaranteed in Serbia (Article 3), as well as the division of power into legislative, executive, and judicial branches (Article 4). The bearer of legislative power is the national assembly, which consists of 250 deputies, while the executive bodies are the president of the republic and the government. When it comes to territorial organization, the Republic of Serbia has two autonomous provinces – the Autonomous Province of Vojvodina and the Autonomous Province of Kosovo and Metohija. AP Vojvodina has its own assembly and the highest legal act – the statute. The Autonomous Province of Kosovo has been under the jurisdiction of the United Nations since 1999 on the basis of UN Security Council Resolution 1244, which was adopted on 10 June 1999. The impact of the market economy on the social and economic position of employees is harmonized through social dialogue between trade unions and employers (Article 82). According to the number of trade union organizations, Serbia is at the very top of Europe (there are currently over 24,000 trade union units). However, the adoption of the amendments to the labor law from 2014, which were opposed by all unions, was done without consulting the workers, who protested several times on the streets of the cities. Although there are several large unions in Serbia, in which smaller ones are integrated, it seems that this is not enough to bring the position of workers to the European level.

According to the Esping-Anderson methodology, the model of the welfare state that is present in Serbia can be classified into hybrid models (a combination of the social democratic, liberal, and conservative models). Namely, some elements of the model were inherited from the socialist system of the former Yugoslavia, while some new ones were introduced by political and economic transformation after the 1990s. Under the exclusively inherited elements of the previous system are those elements that are characteristic of social democratic regimes: universal access to the education and health care system, generous pension systems and the dominant role of the state in providing social services. During the transition, elements of the liberal regime were added to these inherited elements by reducing the amounts set aside for social assistance. The labor market has seen a large share of the secondary sector (Serbia 25 percent, the EU 17 percent) and agriculture (close to 20 percent), as well as high unemployment, which is a consequence of market liberalization immediately after the start of the transition process. In addition, it is very common in Serbia that employed family members receive equal social care for both the youngest and the oldest family members (pensioners). This last is a feature of the conservative model.

Micro-, small-, and medium-sized enterprises (MSME) and entrepreneurs have a significant role to play in all economies, especially in transition and developing countries, which face major challenges in tackling high unemployment and unequal wage distribution. SMEs make a significant contribution to increasing income and creating new jobs on the one hand. On the other hand, they also encourage innovation and development of new technologies and have a direct impact on the level of aggregate demand and investment. In an effort to point out the role and importance of the MSME sector in the growth and development of national economies, in addition to analyzing the participation of this sector in basic macroeconomic indicators, it is necessary to pay special attention to the impact of a large and dynamic SME sector on overall economic performance. one economy.

Thanks to their flexibility, vitality, propensity to undertake innovative and risky ventures, and greater opportunities for specialization, SMEs are superior to large business systems to adapt to consumer demands and dynamic changes in business conditions in the global market.

The largest part of the SME sector is microenterprises that employ up to five people and often only a few family members, while the functioning of large systems takes place in a hierarchy, i.e., a large number of management levels, clear division of labor, formal procedures, and rules for decision-making. The role of SMEs is important at all levels of economic and especially industrial development because these companies represent the backbone of the private sector of each economy, participating with over 90 percent in the total number of economic entities and generating between 50 and 60 percent of total and 40 and 80 percent of employment in production.

9.3 DIGITALIZATION POLICY IN THE COUNTRY

The development of the information society in Serbia began with its accession to the Electronic South Eastern Europe Initiative (eSEE) in 2000. This initiative was launched within the framework of the Second Working Table of the Stability Pact for Southeast Europe, held in October 2000 in Istanbul. The main goals of the Second Working Table are to promote faster economic development based on information and communication technology, integration of Southeast European countries into the European and global market, increasing the global competitiveness of the region, and increasing employment and living standards of the population, i.e., social welfare. By joining the initiative for the spread and acceptance of information and communication technologies in the countries of Southeast Europe, Serbia, together with other countries in the region, began building and developing the information society and, unlike other elements of transition in this domain, there were no delays. However, the bearers of economic and economic development policy in Serbia realized the importance of Information and Communication Technologies late and were not sufficiently aware of the challenges that arise for each economy from the growing global competitiveness of developed countries. Only in 2005, the "National Strategy for EU Accession" envisaged the development of ICT in various sectors, increasing the level of investment in the implementation of ICT and the development of the information society, as well as the harmonization of domestic legislation and legislation with the EU legal framework information society, telecommunications, and media. The importance of ICT and the information society has been recognized only since the second half of 2007, when the knowledge-based economy was set as one of the three pillars of sustainable development (in addition to economic and social issues and the environment) within the "Sustainable

Development Strategy of Serbia." Before that, in 2003, the Republic Institute for Informatics and the Internet was established as a special institution whose competence is the promotion and monitoring of the development of e-government, standardization in the field of ICT, electronic security, and the use of the internet. However, this institution has had and still has an impact only on the technological part of the development of the information society. The economic, legal, and social aspect began to be seen only in the mentioned strategy. The institutional framework for the development of the information society is clearly defined in the "Strategy for the Development of the Information Society in the Republic of Serbia." This strategy is the basic document that defines the framework for ICT policy in Serbia. The strategy was adopted by the Government of the Republic of Serbia in October 2006. In 2010, a new "Strategy for the Development of the Information Society in the Republic of Serbia until 2020" was adopted, which should harmonize the economic development of Serbia with the goals set in the EU economic strategy called "Europe 2020." The process of defining and writing the strategy started in April 2005, which is quite late considering other countries in transition (Slovenia 2001, Macedonia and Croatia 2002, Montenegro 2004). Until 2012, there was a Ministry of Telecommunications and Information Society (MTID) in Serbia, which was a strategic decision of the Serbian government on the road to a digital economy. Along with the Republic Institute for Informatics and the Internet (RZII) and the Republic Agency for Telecommunications (RATEL), these state bodies have constituted a good formal institutional basis for promoting the implementation of the information society and as such could be seen as Serbia's strengths in its development. In addition, the adopted National Strategy for Economic Development and the National Strategy for the Development of the Information Society are of great importance for the transition to the digital economy. However, MTID was abolished in 2012 and the Digital Agenda Directorate at the Ministry of Trade, Tourism, and Telecommunications was formed, which is, among other things, in charge of implementing the provisions taken from the "Information Society Development Strategy in the Republic of Serbia until 2020." However, the adoption of some important development decisions, which primarily relate to supporting the development of companies in the field of information technology, was partly transferred to the Ministry of Finance and Economy, where the position of advisor for competitiveness and knowledge economy was established.

Although Serbia started adopting ICT relatively late, as a result of its belated transition and economic closure during the 1990s, its population shows great interest in using ICT products and services, especially after 2005. The positive side of Serbia is that the population very quickly and spontaneously accepts new information and communication technologies, which is partly due to the activities of various entities (computer training courses, etc.) or the competition itself in certain fields (mobile telephony). Given that after 2005, the ICT infrastructure in Serbia is being repaired and more and more modern equipment is being procured, the acceptance of ICT products from an increasing part of the population creates preconditions for further accelerated development of the information society.

Industry 4.0, the concept of the fourth Industrial Revolution for the next decade, has been accepted worldwide as crucial in the technological and industrial development of modern society. The first IR brought mechanization and steam engines, the second Industrial Revolution mass production, electricity, and measurements, the third Industrial Revolution computers, control, electronics, and automation. The fourth IR brings the integration of classic industrial methods with modern "smart" technologies and the Internet. The technologies that are key when it comes to Industry 4.0 are the Internet of Things, the Industrial Internet of Things, big data, virtual/augmented reality, cloud computing, smart tech, cyber physical systems, 5G, machine learning, etc. All these technologies are already present in commercial everyday devices and consumer electronics, and it is necessary for industry to benefit from the application of these innovations, as it is known that industry always relies on tested and safe, although perhaps outdated, technologies, and that it approaches new solutions reluctantly and with skepticism. These technologies enable a whole new level of control, management, and organization of industry in almost all fields. Decentralization of resources, autonomy of decision-making systems without human assistance, remote monitoring and management, rapid adaptation to new conditions and requirements, connectivity with other systems around the world, greater efficiency and security – these are just some of the reasons why Industry 4.0 is accepted in countries such as Canada, Japan, Germany, Australia, Austria, Switzerland, and the US. Since the initial idea presented in 2011 in Germany, this trend has become generally accepted in the world, so in 2019 Serbia officially joined the technology development strategy by adopting the digital platform for Industry 4.0. For an economy like Serbia's, burdened by the legacy of wars, sanctions, privatization, negligence, and stagnation, this is a unique opportunity to get involved and stay in the company of technology leaders. Therefore, it is of great importance that future generations of engineers are already instructed, educated, and interested in the topic of Industry 4.0, which with its breadth and comprehensiveness enables development in all fields of industry and technology.

The concept of Industry 4.0 includes a broad range of concepts such as robotics, artificial intelligence, nanotechnology, biotechnology, the Internet of things, 3D printing, autonomous vehicles, cloud computing, and big data. As such, they describe organizations and within them the same technology-based production processes and devices that communicate autonomously with each other creating virtual computing environments. Innovations using the latest technologies increase productivity, competitiveness, and generally a market advantage over other branch industries.

In addition, Industry 4.0 implies changes in standardization, new business models, information security, product availability, research, availability of adequate workforce, work processes and changes in the organizations themselves. Industry 4.0 is the current trend of automation and data exchange in manufacturing technologies. This includes cyber-physical systems, the Internet of Things, and computers in the cloud (Kagermann et al., 2013).

Some compare the term *Industrie 4.0* to the fourth Industrial Revolution. However, the latter refers to a systemic transformation that involves the impact on civil society, the management of structures and human identity in addition to the exclusively economic-productive consequences. The first industrial revolution mobilized the mechanization of production by means of water and the power of steam; The second Industrial Revolution introduced electricity into mass production, while the third Industrial Revolution, known as the digital revolution, introduced the use of electronics and IT and further automated production. The term *fourth Industrial Revolution* is based on significant technological developments in the last 75 years, and is ready for academic discussion. The term *Industry 4.0,* on the other hand, focuses on production especially in the current context, and is thus separate from the fourth Industrial Revolution in terms of scale (Kagermann et al., 2013).

History articulates four industrial revolutions:

- The first Industrial Revolution – 1760 to 1840.
- The second Industrial Revolution – 1870 to 1914.
- The third Industrial Revolution – also called the Digital Revolution; 1960–present.
- The fourth Industrial Revolution – an upgrade to the digital revolution; 2011–present.

The term *Industrie 4.0* came to life in 2011 at the Hannover Fair. In October 2012, working groups of the German federal government presented a series of recommendations for the implementation of Industry 4.0. The members of the working group were recognized as the founders and driving force of Industry 4.0 (Kagermann et al., 2013).

9.4 METHODOLOGY

Looking at the EU, the move towards a digital single market is a challenge for SMEs, which with a turnover of €23 million form the backbone of the economy. Current estimates show that more than 40 percent of them still do not take advantage of opportunities related to digitization, because they are not sure what will be the effects of the value chain, although it turned out that companies that use digital technology grow two to three times faster, are more productive, and employ more workers (Euractiv, 2015). In order to achieve this, governments, industry, NGOs and other decision makers from 22 EU member states have launched an "e-Skills for Jobs in 2015" campaign, signing a declaration in Riga. The declaration includes ten principles, including more and better investment in digital technologies and e-skills, the fight against youth unemployment in Europe through digital opportunities, and the promotion of e-leadership at the management level in European companies. Unions are wary of change, as job-related questions are asked: what jobs could be compromised, what implications may there be in terms of controlling performance, accountability, and stress? (Smit et al., 2016).

It is undeniable that the implementation of Industry 4.0 requires a wide range of skills required throughout the value chain at the operational and support level, from infrastructure through system design, modeling and production process management to human interaction skills. The convergence of IT systems, manufacturing, technologies, and automation software requires the development of completely new approaches to the education and training of IT professionals (Smit et al., 2016).

And here comes the term *STEM*, which actually defines the areas from which the new experts of Industry 4.0 will emerge – science, technology, engineering, and mathematics. The advantage will be achieved by those who successfully connect several areas, i.e., gain expertise from several of them.

According to a recent study by the European Commission, digital know-how to support the Digital Single Market is short on expert resources. Estimates of the commission show that by 2022, Europe could lose about 756,000 IT professionals. On 18 April 2016, the European Commission published a "Communication on the Digitalisation of European Industry," which introduced a series of meaningful measures within the Digital Single Market in the form of technology and public service modernization arrangements. Part of the press release is dedicated to digital skills. In particular, it calls for human capital to be ready for digital transformation with the necessary skills.

Industry 4.0's vision is that everything will be networked through Cyber-Physical Systems – people, things, processes, services, and data. Production in Industry 4.0 will be more flexible and faster. The data can be made available to everyone involved in the process, and in real time. Increase of interrelated data should bring greater efficiency and improved productivity since resources would be used efficiently. This should ultimately lead to improved sustainability.

Industry 4.0 is predicted to be fully implemented only in 2025. In order to launch an initiative on the road to implementation, preconditions must be met, including technological infrastructure, work systems with the required level of security, and people who will design them and who have a vision of developing new ways of working. There are also concerns that the competition will do the same or even better, and potentially even less expensively. These are risks that are justified and that can influence the policies of public administration and companies to orient towards Industry 4.0 (Smit et al., 2016).

9.5 INDUSTRY 4.0 BARRIERS

One of the main challenges in the implementation of Industry 4.0 is the lack of skilled labor and the requirement of retraining staff to adapt to the changed circumstances. In the future, new ways of working will be needed, which can have positive and negative effects on employees. Changed working conditions can lead to conflicts in business organizations.

Numerous sources (e.g., Erol et al., 2016; Kiel et al., 2017; Müller & Voigt, 2016) suggest that the lack of financial resources as a major obstacle to the implementation of Industry 4.0, as well as a low degree of standardization, poor understanding of integration and concerns about data security.

The development of production systems also significantly affects the risk of fragility, creating further uncertainties in the ecosystem. Successful integration of components, tools, and methods requires the development of a flexible interface, as the synchronization of different languages, technologies, and methods can lead to significant challenges. The reliability and stability of the system must also be ensured, and this is a crucial factor in machine-to-machine communication.

Within these constraints, SMEs in particular face certain barriers to participating in the Industry 4.0 supply chain (Smit et al., 2016):

- Lack of awareness of advanced technologies and potential advantages of applying them in production processes.
- The ability to purchase the necessary technology and invest sufficiently in research and development where current technology is not available.
- Capacity to launch pilot projects and potentially limited access to facilities for testing advanced solutions.
- The availability of qualified and specialized professionals needed for the integration and use of advanced machine tools, but also the ability of SMEs to attract such a skilled workforce.
- Large companies can use their market position to be the first to test and patent a product in the first test.
- Industry 4.0 may result in internationalization of production – an easier task for large corporations than for SMEs, which may result in increased dependence of SMEs on larger companies.

9.6 INDUSTRY 4.0 DRIVERS

Current changes at the global level have led to the networking of society, affecting both business and private life. These challenges include a reduced number of labor due to population decline and aging of the society. They can also be addressed by the development and application of new technologies. Growing levels of competition have made it crucial for companies to increase their innovation capacity and productivity and reduce their market penetration (Kagermann et al., 2013; Lasi et al., 2014).

Investments in new digital technologies enable companies to improve their comparability advantage and create a decisive advantage over competitors. The change is also forced by the reduction of product life cycles, changing expectations, and needs of consumers and markets over time becoming more heterogeneous. By improving productivity, the quality of workmanship can be significantly increased and waste reduced. Significant improvements can also be achieved in energy efficiency and Industry 4.0 can positively affect the production of environmentally sustainable production, the development of green products, production processes, and supply chain management. Therefore, companies can rely on Industry 4.0 for an increase in sales volume, to achieve significant cost savings, and to provide radical performance improvements at the micro level. Also, the collection and processing of production data from the field supports other benefits, e.g., faster decision-making and knowledge management support. Industry 4.0 technologies help manage planning and channel scheduling, capacity utilization, maintenance, and energy management. Industry 4.0 can also lead to significant changes in existing business models, providing new ways to create value. These changes are expected to result in the transformation of traditional value chains and create entirely new business models that enable a higher level of consumer involvement (Kagermann et al., 2013; Müller et al., 2018).

Industry 4.0 affects the three elements of manufacturing SMEs: creating value, capturing value, and offer value. This can lead to a change in customer relationships and increase product innovation and service design. Industry 4.0 can therefore be defined as a basic pillar in the future competitiveness of manufacturing companies. Firms, however, will face challenges in implementing it.

9.7 INDUSTRY 4.0 OPPORTUNITIES

Most companies today recognize the likely impact of Industry 4.0. However, SMEs are generally less prepared for the new technology. After the discussions, it is clear that there are opportunities brought by the new Industrial Revolution, but there are also challenges. Research studies focus on four categories, which can be simultaneously placed in the domain of opportunities and challenges, namely the transformation of the manufacturing sector, new aspects of operational efficiency, changing business models, and laying the foundations for digital transformation.

Modern and developed countries, either following the initiative of large corporations or based on government decisions, are working intensively and systematically on the transformation towards Industry 4.0. In fact, a good part of them are already in the digital transformation. Which cannot be said for Serbia. In Croatia's industrial framework, there is currently no initiative to make the transition towards a new industrial paradigm, but there is no understanding of the need to reform the education system in a way to make a step towards generating educated staff in the future.

Circumstances that have taken place in the last 20 or more years have not been in favor of the development of industrial production; on the contrary, it has been drastically reduced. However, it should be emphasized that the development phase dating from the self-management model of production management could not remain competitive in the long run internationally, for several reasons. One is certainly political, which is not part of these considerations, the other is the technological gap that has occurred with Western production systems, and the extremely low production costs of Eastern emerging economies. In such an environment, the only way to occupy a certain space in regional and international frameworks is to be special, innovative, technologically advanced, and ultimately competitive. Industry 4.0 is exactly the opportunity to be seized as soon as possible. The stimulus wave should come from two directions – a clear, profiled, and unquestionable state economic policy and the previously mentioned education system. The STEM initiative exists, albeit not by the relevant institutions but by individuals who have recognized the moment in which the last train needs to be boarded in order to catch a connection with the developed world.

According to Roland Berger's research on the readiness index for Industry 4.0, Serbia belongs to the group of hesitant countries, and they are characterized by a very low readiness index. It is clear that this is not a good indicator and that this issue needs to be addressed systematically. There is no need to cultivate hopes that industrial production will return as it was in history, primarily heavy industry; we should, rather, focus on technologically advanced industries that require skilled labor, have relatively high incomes, and easily find their position in the market.

9.8 CONCLUSION

The times in which humanity finds itself are subject to changes that are taking place faster than ever before in human history. Especially in the segment of social and economic frameworks. They can be attributed to the processes of globalization, from which they certainly arose and which accelerated them, but also to the technological development, which, on the basis of new scientific knowledge, experienced a real revolution, the fourth in a row.

The development of IT technologies in the last 20 years has further emphasized the digital transformation of industrial production as a direction that must be taken. The internet as a platform has created an incredibly extensive network and the possibility of mutual communication to such an extent that everything that surrounds people is connected, or, rather so little is left that is not in the global chain of communication. Concepts like cloud computing, the Internet of Things, 3D printing, big data, and the like have become the foundation of today's modern, industrial world. New value chains, innovative products, product customer interaction with the production process, networking of all factors of production, data as a core value – all these are terms that are the alphabet of Industry 4.0.

The results of the research showed that the term *Industry 4.0* is not widespread in the Serbian public, much less present in industrial production. Analyses have shown that from the position of Croatian industry and the thinking of citizens, it is about the future, which, although it has potential, is viewed with a certain distrust and misunderstanding. If viewed from the position of leading economies, it is a present that is already in a deep transition towards a complete fourth Industrial Revolution.

REFERENCES

Erol, S., Jäger, A., Hold, P., Ott, K., & Sihn, W. (2016). Tangible Industry 4.0: A scenario-based approach to learning for the future of production. *Procedia CIRP*, 54, 13–18.

Euractiv (2015). *EU 'racing to catch up' with digital single market plan*. Available at: www.euractiv.com/section/digital/news/eu-racing-to-catch-up-with-digital-single-market-plan/

Kagermann, H., Wahlster, W., & Helbig, J. (2013). Recommendations for implementing the strategic initiative INDUSTRIE 4.0, in: Final report of the Industrie 4.0 Working Group, Acatech, Frankfurt am Main, Germany.

Kiel, D., Müller, J. M., Arnold, C., & Voigt, K. I. (2020). Sustainable industrial value creation: Benefits and challenges of industry 4.0. *International Journal of Innovation Management*, 21(8), 231–270.

Lasi, H., Fettke, P., Kemper, H. G., Feld, T., & Hoffmann, M. (2014). Industrie 4.0. *Business & Information Systems Engineering*, 56(4), 261–264.

Müller, J. M., Buliga, O., & Voigt, K. I. (2018). Fortune favors the prepared: How SMEs approach business model innovations in Industry 4.0. *Technological Forecasting and Social Change*, 132, 2–17.

Müller, J., & Voigt, K. I. (2016). Industrie 4.0 für kleine und mittlere Unternehmen. *Productivity Management*, 3, 28–30.

Smit, J. Kreutzer, S., Moeller, C. & Carlberg, M. (2016). *Industry 4.0*. European Commission. Available at: https://www.europarl.europa.eu/RegData/etudes/STUD/2016/570007/IPOL_STU(2016)570007_EN.pdf

10 Birthplace of the First Industrial Revolution
Fertile Ground for SMEs in Its Fourth Incarnation

Shaden Jaradat

CONTENTS

10.1 INTRODUCTION

The first Industrial Revolution was born in England in the late 1700s and, for much of the time since, manufacturing has been part of the fabric of the UK's economy, society, and geopolitics. Currently, manufacturing contributes 11 percent of gross value added, with goods accounting for 53 percent of the UK's total exports. It employs approximately 2.7 million people, and accounts for nearly 65 percent of Research and Development (R&D) investment. However, UK manufacturing industries have been facing significant – and growing – challenges. Barring one year since 2013, the UK remained the 9th largest manufacturer in the world (Make UK 2020).

Investment in industrial digital technologies (IDTs) has started ramping up relatively recently. The skills gap across the workforce is often cited as an inhibitor of competitiveness, while the slow adoption of IDTs continues to hold back the growth of a large number of UK small and medium enterprises (SMEs). It is worth noting that the UK government defines an SME as a firm meeting two out of three criteria: staff of fewer than 250; turnover of less than £25 million; or gross assets of less than £12.5 million. According to 2020 data by the UK's Department for Business, Energy & Industrial Strategy (BEIS), 99.3 percent of SMEs were small or very small businesses, including 4.4 million sole traders.

DOI: 10.1201/9781003165880-10

Moreover, despite numerous research and innovation funding schemes aimed at bringing UK universities to the forefront of industrially applicable scientific advancements, the UK manufacturing ecosystem is still finding it difficult to push many of these advancements higher up the technology readiness levels.

However, alongside these challenges, a myriad of opportunities are also available. The rate at which enabling technologies are evolving, such as the increased potential of advanced materials to support better productivity and smarter processes, is expanding the arsenal at the disposal of innovators. Customers' growing desire for zero-defect, low-emission, sustainable production has ensured manufacturers, researchers and governments alike are keen to embed these targets in their processes, innovation, and policies. The proximity of UK manufacturing companies to urban areas, where 83 percent of the UK population live (World Bank 2020), strengthens the potential for localized supply chains. Furthermore, in the context of manufacturing, the COVID-19 pandemic could propel an acceleration of the adoption of Industry 4.0 technologies as a means towards greater resilience against similar disruption in the future.

This chapter will describe the digitalization landscape and policies in the UK, focusing on the dominance of the industrial strategy in the past ten years, the growing political will and societal desire to homogenize the UK's regional economy, and the central role of research institutions. Barriers to Industry 4.0 impacting UK SMEs will be discussed, including the lack of trained workforce, attitude towards adoption, and user trust. Several avenues with the potential to unlock opportunities to address these challenges will be discussed.

10.2 APPROACH AND METHODOLOGY

This chapter will rely on two aspects to develop its narrative. Firstly: The UK's manufacturing-related research and innovation landscape will be used as a proxy indicator of national, regional, thematic, or sectoral opportunities and barriers relevant to Industry 4.0, especially those where SMEs have an important role. This is because, on the one hand, the UK's industrial strategy was only published in November 2017 (BEIS 2017) and did not have a predecessor; therefore, the success of its output cannot yet be fully assessed. Another, more established compass is required. On the other hand, research policy has been based on more solid foundations and clearer policy for a longer period. For example, the government has been committed to increasing UK investment in R&D to 2.4 percent of GDP by 2027 and £22 billion of public funding per year by 2024/25 (BEIS 2017). Moreover, research priorities have been influenced by business through, for example, the inclusion of route-to-market impact as one of the measures used in the Research Excellence Framework which is used to evaluate British higher education (Dowling 2015).

Secondly: The region of Greater Manchester (referred to henceforth as Manchester) will be used as a case study illustrating how national policies cascade to the regional level. The rationale for examining the case of Manchester is twofold: (1) The devolution of some powers from central government to some city regions across England started in 2014. The devolution to Manchester has gone further than any other region in England outside London, and included spatial planning, skills, health, waste management, transport, housing, security, and justice (Local Government Association website 2021). The city is aware of the challenges facing its own growth (skills and pay gap, and lagging in productivity) and its world-class strengths in advanced materials, health innovation, manufacturing, digital and creative, energy and industrial biotechnology (GMIPR 2019), making it an ideal testbed for policies, initiatives and platforms that aim at empowering its manufacturing base, particularly SMEs. (2) While the UK government has set a target of achieving net zero emissions by 2050, together with a 10-point plan for a Green Industrial Revolution (UK Government 2020 (1)), Manchester set out a more ambitious target of net zero carbon by 2038 (Manchester Zero Carbon Framework 2020). Crucially, Industry 4.0 is widely perceived by specialists, strategists, academics, and large manufacturers to enable and accelerate the realization of greener economy and manufacturing. Given that 99 percent of UK firms are SMEs (BEIS 2020), it would be hard to envisage achieving these targets without significant and early buy-in from them, not just in committing to the government's legal requirements of net zero, but by adopting the means to achieving them through Industry 4.0.

This chapter will primarily utilize information and data from grey literature published by national/local governments, non-governmental organizations and think tanks concerned with manufacturing, productivity, and IDTs, as well as sector-wide representatives and trade associations. The overall narrative was guided by an amalgamation of (1) SWOT analysis coordinated by the author across an interdisciplinary group of University of Manchester (UoM) researchers representing science and engineering, social sciences and life sciences in 2018–2019, and (2) a qualitative understanding of the challenges and opportunities around adoption of Industry 4.0 technologies by SMEs was gleaned from the author's discussions with businesses and trade associations researching these aspects.

10.3 DIGITALIZATION LANDSCAPE AND POLICIES IN THE UNITED KINGDOM

There are three central pillars that underpin the UK's digitalization landscape, which will be denoted here as short-term, long-term, and inherent.

1. *Short-term – Industrial Strategy and Made Smarter:* The "Industrial Strategy: Building a Britain fit for the future" (BEIS 2017) white paper was published as a roadmap for a more productive economy. It defined four Grand Challenges where innovation and investment will position the UK at the forefront of the industries of the future: Artificial Intelligence and Data Economy; Ageing Society; Clean Growth; and Future of Mobility of People, Goods and Services.

In tandem, the UK government's growing attention to the role of IDTs was demonstrated through commissioning the Made Smarter Review (2017) and led by Jürgen Maier, then CEO of Siemens UK and an honorary professor at UoM. It concluded that the benefit to the UK economy of adopting IDTs over the next decade could reach £455 billion, linked with growth in manufacturing of up to 3 percent per annum, the creation of 175,000 jobs and a 4.5 percent reduction in carbon emissions. The Review became a landmark in UK government policy towards IDTs, and triggered a propagation of national and regional initiatives with Industry 4.0 at their heart. Its key recommendations are: (1) Leveraging the UK's innovation capabilities to accelerate commercialization of IDTs, (2) Stronger leadership, vision and branding of the UK's world-class capabilities in IDTs, (3) Addressing barriers in the way of adoption of IDTs particularly by SMEs, and (4) Upskilling a million industrial workers to enable IDTs to be successfully exploited.

2. *Long-term – The North-South Divide:* The regional imbalance manifesting itself in England's North-South Divide is a longstanding notion, and has induced a spectrum of analysis ranging from considering it a historic perception (Hughes and Atkinson 2018) to an ongoing and growing consequence of the deindustrialization of the 1980s (Beatty and Fothergill 2016). Either way, the rates of private and public job creation since the 2008 economy crash have been slower in the northern regions of England than the South and London in particular (SPERI 2015). The government's attempts to address this divide in the past decade has assumed various incarnations, most notably, the Northern Powerhouse strategy (Northern Powerhouse 2016) which represented the government's vision for a "super-connected, globally-competitive northern economy with a flourishing private sector, a highly-skilled population, and world-renowned civic and business leadership." The strategy identified manufacturing, pharmaceuticals, energy and digital as four sectors in which the northern region has significant strengths. It is also worth noting that the Made Smarter Review highlighted the need for digital hubs and recommended the establishment of one in the North West of England, leading to the creation of the Made Smarter NW Pilot (www.madesmarter.uk) to help accelerate the development and diffusion of IDTs through focused support to SMEs. Since late 2019, these visions have been broadly incorporated in the government's "Levelling-up Agenda" which is firmly taking center stage in UK policies and is being emphasized as a result of COVID-19, as illustrated by the 2020 Budget (Treasury 2020 (1)). In parallel, the 2050 net zero target is being embedded in these policies, with a green industrial revolution being championed as the means to interlink – and deliver – them. The impact of these policies is beginning to materialize in national strategies such as the National Infrastructure Strategy – "Fairer, Faster, Greener" (Treasury 2020 (2)) – which sets out plans to transform UK infrastructure in order to level up the country and achieve net zero by 2050, while putting innovation, new technology, and private investment, at the heart of the government's approach.

Being one of Europe's fastest growing cities (Deloitte 2021), Manchester is poised to continue playing a central role in the underlying drive to bridge the UK's North-South divide. In response to the national Industrial Strategy, Manchester formulated a Local Industrial Strategy (LIS) (UK Government 2019 (1)), which recognizes that "The city-region's manufacturing industry, which employs 110,000 people and generates £8bn of economic output each year, is being transformed by the fourth Industrial Revolution." The LIS sets out plans that are highly relevant to Industry 4.0, notably: positioning Manchester as a world leader for innovative firms to trial, develop and adopt advanced materials in manufacturing; maximizing growing assets in cyber security, enabling the digitalization of all sectors and capitalizing on the links between digital and creative industries; and ensuring that the education, skills, and employment systems enable employers to access the required skills.

3. *Inherent: Research and Innovation:* Research organizations in general, and universities in particular, hold a special and highly revered place in British manufacturing psyche, which may be attributed to the following factors among others: (1) UK universities were the birthplaces of many historic "firsts" that created the foundations for entire sectors. Manchester-based examples include: first splitting of the atom (Rutherford), first stored-program computer (Williams, Kilburn, Tootill), first textbook defining the discipline of chemical engineering (Davis), and graphene (Geim, Novoselov). (2) Today, universities remain the go-to powerhouses to help drive forward major projects and initiatives that require a great deal of innovation and novelty. The most vivid and recent example is the only successful COVID-19 vaccine to date that carries the name of a university (Oxford/AstraZeneca). (3) Many world-leading companies have strong, long-standing relations with UK universities manifested in joint research centers and training programs. For example: Rolls Royce has University Technology Centre partnerships with the universities of Birmingham, Bristol, Cambridge, Cranfield, Imperial College, Loughborough, Manchester, Nottingham, Oxford, Sheffield, Southampton, Strathclyde, Surrey and Swansea (Rolls Royce 2021); Siemens' Industry 4.0 "Connected Curriculum" has been adopted by the universities of Sheffield, Liverpool John Moores, Middlesex, Teesside, Coventry, Exeter, Salford, and Manchester Metropolitan (Siemens 2019); Tata Steel has an R&D facility at the University of Warwick; the Advanced Machinery & Productivity Institute (www.ampi.org.uk), which will be a hub for innovation in machinery technologies and supply chains for precision components, control systems, advanced software development, Industry 4.0, digital systems and intelligent networks, is led by a consortium consisting of three Manchester universities alongside various industrial partners. (4) The Catapult Network (catapult.org.uk) draws upon the UK's academic community to help businesses in transforming ideas into valuable products and services. Catapults can be viewed as an indicator of where some of today's innovation in technology can contribute to future growth. Relevant catapults are: Compound Semiconductor Applications; Digital; Energy Systems; High Value Manufacturing; Offshore Renewable Energy; Satellite Applications.

UK Research Councils have, over many years, created an environment and tradition of de-risking industrial investment in R&D and enabling partnerships between universities and companies. The Research Council that is most geared towards supporting businesses – Innovate UK – follows a model whereby the level of de-risking offered to companies is dependent on their size. For example, large companies may get funding up to 25 percent of the total costs of an experimental development project that is nearer to market, whereas this recovery rate can be as high as 45 percent for micro or small organizations. Feasibility studies and lower TRL projects can offer up to 50 percent and 75 percent recovery for similar organization sizes, respectively (Innovate UK 2021).

More broadly, the current landscape is largely shaped by the government's target to increase total investment in research and development to 2.4 percent of GDP by 2027. The suite of research funding schemes supporting the Industrial Strategy is the Industrial Strategy Challenge Fund (ISCF). It has been backed by £2.6 billion of public money and £3 billion in matched funding from the private sector, covering over 20 themes under the banners of the aforementioned four Grand Challenges (UKRI 2021 (1)). One of these themes is Manufacturing Made Smarter, which encapsulates much of the innovation-related recommendations of the Made Smarter Review. It comprise a £147 million investment, matched by a minimum of £147 million from industry, aiming to deliver a resilient, flexible, more productive, and environmentally sustainable UK manufacturing sector. While the collection of initiatives under the Made Smarter banner has come to embody the UK government's brand of Industry 4.0, various other themes under ISCF are strongly underpinned by Industry 4.0 technologies. For example: industrial decarbonization, transforming construction, smart sustainable plastic packaging, transforming foundation industries, precision medicine, robots for a safer world, and next generation services.

It is also worth considering the infrastructure that supports UK research and innovation specifically in physical science domains, such as: engineering, energy and environment, and computational and e-infrastructure (UKRI 2020 (1)). Research Councils recognize that these infrastructures currently have their own challenges. Unsurprisingly they, too, will rely heavily on Industry 4.0 technologies in order to remain relevant. For example: (1) The long lifetimes of some of these infrastructures need to take into consideration continuous technical advancements throughout their lifecycles, (2) although energy, environment, low-carbon and integration of renewable sources are key priorities, the UK has relatively modest capabilities in solar, bioenergy, power distribution, hydrogen and fuel cells, and energy system transition, and they tend to be mainly based in universities, and (3) between 70–80 percent of the infrastructure in these sectors envisage e-infrastructure and data becoming more relevant to them in the next decade. Industry 4.0 technologies that are particularly crucial for the sustainability and evolution of the infrastructure include: imaging & characterization; robotics & automation; data analysis; advanced materials; monitoring of whole environments; simulation and prediction; data analytics; machine learning and AI; and cloud, cyber security and authentication (UKRI 2020 (2)).

10.4 INDUSTRY 4.0 BARRIERS

Based on an extensive set of studies, surveys and insights, the Made Smarter Review identified the following gaps in the way of achieving widespread utilization of IDTs: absence of a leadership-level narrative and strategy; lack of a coordinated industrial R&D environment to capitalize on the UK's innovation strengths, which is manifested in slow commercialization; lack of adoption, particularly among SMEs; and the skills shortage.

Some of these gaps have in recent years seen significant efforts to bridge them, with tangible outcomes already noticed. In relation to leadership, the Made Smarter Review itself has established the foundation for a UK brand of Industry 4.0, instigating a variety of coordinated programs, and feeding into a raft of initiatives and national or regional strategies. In effect, Made Smarter has triggered a movement that is shaping a roadmap to address the leadership gap. As for commercialization, a recent government-commissioned report (BEIS and RSM PACEC Ltd 2018) concluded that commercialization of university intellectual property is showing marked improvements, thanks to a combination of the growth in the number of university-established commercialization companies and new policy developments (including increased funding) to support and incentivize the focus on research impact. With regards to adoption, COVID-19 has highlighted the need for recovery, diversification and resilience, with faster adoption of Industry 4.0 being key to accelerating the realization of these goals (CIIP 2020). But is the mere *desire* enough to push the resource-limited SMEs over the adoption line? And what about customer and user trust? Lastly, shortage of a skilled workforce is a prolonged and cross-generational challenge which requires persistent efforts from a wide spectrum of stakeholders.

It is those latter barriers, namely, upskilling and attitudes towards adoption, which will be the elaborated on in the following:

1. Even prior to the Made Smarter Review and the Industrial Strategy, the lack of trained and skillful graduates in engineering and technology has been widely identified as a major challenge in the way of unlocking the benefits of Industry 4.0 across UK manufacturing sectors (UKCES 2015; IET 2015). This is exacerbated by the expectation that UK manufacturing companies will need ever more staff with IT skills and production-related technical skills – up to 50 percent and 60 percent, respectively (EEF 2016). Interestingly, during the past ten years, entries into engineering and technology undergraduate degrees and postgraduate research degrees have increased by 6 percent and 10 percent respectively, while entries into postgraduate taught degrees have fallen by 5 percent (Engineering UK 2020). This suggests that, rather than it being an issue of low demand from students for engineering and technology degrees, the skills shortage could be a consequence – at least in part – of a higher education output that is not Industry 4.0-ready. This challenge adds further complexity to the existing divide in terms of skills and talent. For example, a higher proportion of Manchester's workforce is employed in science, research, engineering and technology professions than the England average, but a significant share of the local population continues to have no qualifications at all (UoM 2016).
2. Adoption is a two-way barrier, which could be attributed – as will be argued next – to attitudes. Firstly, uptake from SMEs is not commensurate with the perceived opportunities of Industry 4.0. To illustrate, let us consider AI-optimized manufacturing, an area where the UK investment is reported to be the largest in Europe, nearing £1 billion in 2018 (Diffey 2019). Although 58.2 percent of surveyed manufacturers are aware of the benefits that AI can bring to their business, only 23 percent of them are currently using it (During 2019). Some of the reasons for the low uptake, from the SMEs' point of view, can be gleaned from a 2018 survey of SMEs in Manchester which showed that 84 percent of them felt the cost was the biggest barrier to adoption, 80 percent cited skills shortages and 90 percent said workforce upskilling/reskilling was a priority (Comrie 2018). It is possible that investing in Industry 4.0 enablers (be it the technology infrastructure or human aspects) is not yet perceived as cost-effective by a large proportion of SMEs. Secondly, user and regulator trust are potentially an equally important part of this barrier that may dampen industry's appetite for adoption, given that safety must not be compromised as the use of technologies such as AI and machine learning become more prevalent (Jarrahi 2018). This could be illustrated using three examples inspired by

> the author's discussions with companies in the sectors of maritime, bio-simulation and driverless vehicles: (1) Despite the marked advances in additive manufacturing technologies and the growth in their use in applications such as bioengineering or light-weighting (Wohlers 2021), the ability to utilize them in extreme environments and safety-critical conditions such as deep-sea operations remains far underdeveloped. (2) Model-informed drug design offers a an increasingly reliable method for enabling precision dosing of medicines in diverse populations where clinical trials are not possible due to underlying health conditions (Certara 2016). However, meeting the high standards required by medical regulating authorities is imperative before this method of drug design and delivery gains traction with patients and practitioners alike. (3) Despite driverless cars being an important element of the UK government's emphasis on the "future of mobility" (BEIS 2017), there is a significant risk from unclassed parts – which have never been tested or approved in a driverless car – being put on the roads. The level of certification and traceability that exist in the autonomous vehicles sector remains far behind those achieved by other sectors such as aviation.

Notably, it could be Industry 4.0 that would also provide the solutions to the challenges we have noted. For example, digital twins of additively manufactured components, patient organs or driverless car factories may be able to achieve the desired trust levels by manufacturers, regulators, and users. ISCF signaled a recognition of this by launching the "Robots for a safer world" challenge (UKRI 2021 (2)) with a £112 million investment focusing on innovation in advanced robotics and autonomous systems to create a safer working world, including extreme and challenging environments, as well as addressing new needs arising from the COVID-19 pandemic (e.g., robotic sanitizing of vital facilities).

10.5 INDUSTRY 4.0 DRIVERS IN THE UK

Earlier in this chapter, the Industrial Strategy (and its Industry 4.0 façade), and the UK's levelling up agenda were discussed. These could be viewed as policy-intense drivers of Industry 4.0, and this is especially the case in relation to SMEs: Firstly, SMEs are at the heart of the Made Smarter initiatives and, secondly, the vision for achieving the levelling-up agenda post–COVID-19 – Build Back Better (Treasury 2021) – aims to incentivize 100,000 SMEs to adopt productivity-enhancing digital technology.

At the time of publication, the UK is at important geo-political and socio-economic junctions, the backdrop of which consists of: an emphasis on the need for more off-shoring due to Brexit (SWMAS 2020), and a sense of urgency in driving forward the levelling-up agenda which already projected itself in aspects such as the R&D roadmap (UK Government 2020 (2)). Then, the COVID-19 pandemic struck and, although it often eclipsed the two drivers, much of the commentary about the post-COVID recovery followed a narrative that is favorable to Industry 4.0. Rather than bringing manufacturing and supply chains back to the previous normal, the "new normal" emerging from the pandemic would be shaped through an acceleration towards the Industry 4.0 paradigm. In other words, Industry 4.0 will continue to underpin the vision of the future of manufacturing and its implications on society and the economy, only the need to accelerate it has been amplified. As we view the UK's Industry 4.0 landscape through the prism of these challenges, we could group the drivers for the vision of the future into three categories, with examples for each one of them in the following:

1. *Technological drivers:* (1) From utilizing graphene through the innovation and scale-up pioneered by the Graphene@Manchester innovation cluster (www.graphene.manchester.ac.uk) to being at the frontiers of additive manufacturing (UKAMSG 2017), the UK is a world-leading power in advanced materials innovation that could play a significant role in revolutionizing manufacturing. (2) Already no stranger to the deployment of Industry 4.0 prior to COVID-19, the pharmaceutical sector is now indisputably one of the most vital globally – not only because of the speed and efficacy in which the first generation of COVID vaccines were designed and manufactured, but also because of the expectation that a greater reliance on Industry 4.0 technologies is needed in order to ensure future generations of the vaccine (and the speed at which they are incorporated into production lines and supply chains) keeps up with COVID-19 mutations. (3) The rapid increase in computational power continues to unlock novel opportunities across more industrial sectors. UKRI recognized that the UK ranked third for AI research & innovation, but 11th in ability to realize the impact from it, and are therefore proposing a raft of measures to capitalize on these capabilities by focusing on application-driven research (UKRI 2021 (3)).
2. *Economic drivers:* (1) Throughout the political discussions around Brexit since 2016, the argument in favor of a clean break between the UK and the EU's free market while retaining "frictionless trade" often referred to smart borders (Karlsson 2018). Broadly, this would also include visible supply chains, traceability of products from origin to destination, and smart assessment of conformity. Indeed, based on the UK-EU deal that transpired, the borders between Northern Ireland and the Republic of Ireland, and between Northern Ireland and the rest of the UK, became trading and political fault lines with ample opportunity for smart technology to step in and remedy. (2) Socially responsible and ethical production, supply and consumption are becoming central to corporate policy and consumer habits (Coop 2019), and can be propelled through mass-personalized customer engagement. (3) Global collaboration and competition alike are important drivers of Industry 4.0, and are aptly exemplified by the world-wide production and supply of the first generation of COVID-19 vaccines. The remarkable reduction of the timescale of producing a vaccine (from several years historically, to just under 12 months in the case of several COVID-19 vaccines) could not have happened without the use of key Industry 4.0 technologies connecting sequencing and clinical data globally. Furthermore, would a better digital connection between government, manufacturers and suppliers have prevented what became known as "vaccine nationalism" (WHO 2021)?

3. *Social and environmental drivers:* (1) Addressing climate change is a multidimensional endeavor for the UK. Its commitment to honoring the Paris agreement (www.unfccc.int) of achieving a carbon-neutral world by mid-century was illustrated by becoming the first major economy to enshrine in law a new target requiring the UK to bring all greenhouse gas emissions to net zero by 2050 (more ambitiously, Manchester has committed to a similar target by 2038). Being host to COP26 (www.ukcop26.org), there is a particular desire for the UK to lead by example. The opportunity for businesses here is to ensure their manufacturing processes would produce more with less resources, which will require capitalizing on research and innovation in circularity, sustainability, and Life Cycle Assessment. (2) The UK has a strong and long-standing tradition of preserving green buffers between towns and between towns and countryside, referred to as Green Belts (Garton and Barton 2020). Combined with a growing innovative and "Industry 4.0 – ready" urban population, small and micro manufacturing facilities will be created close to end-consumers and within the confines – and benefits – of cities. With more of the UK's population living in urban areas than any other OECD country (World Bank 2020), many of those urban manufacturers will focus on personalized and seasonal manufacturing. (3) To help realize the aforementioned social and environmental drivers, a cross-sectoral symbiosis approach to manufacturing will need to be developed to ensure that factories across multiple manufacturing sectors transform waste streams to resource flows.

These drivers are propelled by Industry 4.0 enablers that encompass the technologies underpinning them as well as how they work and interact with people and organizations. These enablers can be grouped into the following three domains, each illustrated by two exemplar activities or technologies of growing vitality in the UK and relevance to SMEs:

1. Human-centric enablers: (1) One of the primary aims of the UK's National Cyber Security Centre (NCSC) is to support industry and SMEs. The national focus is also reflected regionally – for example, the Greater Manchester Cyber Foundry is a European funded business support project aiming to combat cyber threats in the region by drawing upon expertise from universities in Manchester. (2) The challenges around training and upskilling, and various calls for addressing them, have been referred to throughout this chapter, and there are indications that they have become an important element of government investment and public-private partnerships. For example, in its initial response to the independent Post-18 Education and Funding Review Panel (Augar 2019), the government committed to a Lifelong Loan Entitlement from 2025, enabling people to "access funding across higher and further education throughout their lifetime" as well as moving towards "modularisation of higher education in order to provide a truly flexible system that provides more opportunity for upskilling throughout people's careers" (DfE 2021). Another example

is the £37-million government investment to establish a new institute in Manchester to boost research into increasing productivity and wages and supporting economic recovery. Based in the University of Manchester, the Productivity Institute's focus areas include understanding the supply and demand for labor and skills as companies implement new technologies (BEIS Press Release 2020 (1)).

2. Data-centric: (1) UKRI's investment portfolio in AI is in the region of £1 billion including: £135 million in people and skills such as AI centers for doctoral training, £530 million in research and innovation, and £410 million in strategic investments such as the Alan Turing Institute (www.Turing.ac.uk) and Hartree Centre (www.hartree.stfc.ac.uk). (2) One of the main drivers for Computational Fluid Dynamics (CFD) is the need to decrease product development time, particularly in vital manufacturing sectors to the UK such as chemicals, pharmaceuticals, and plastics. Opportunities for businesses leveraging government investment to address challenges in scaling up of manufacturing processes for formulated products are exemplified by the Centre in Advanced Fluid Engineering for Digital Manufacturing (www.CAFE4DM.manchester.ac.uk) – a UKRI supported research hub in partnership between Unilever and a number of UK universities.
3. Machine-centric: (1) Automation and robotics: In its 2019 report, the UK Parliament's Business, Energy and Industrial Strategy Committee concluded that the problem for the UK labor market and economy is "not that we have too many robots in the workplace, but that we have too few," quoting that, in 2015 the UK had just 10 robots for every million hours worked, compared with 167 in Japan (HoC BEIS Committee 2019). The government's response to the report (UK Government 2020 (3)) lists a myriad of robotics and autonomous systems investments through UKRI, reinforcing the central narrative of this chapter that the UK's research landscape is a strong indicator of its Industry 4.0 priorities. (2) Additive Manufacturing is a technology of increasing relevance (Wohlers 2021), with major players in important sectors becoming early investors in it, including nuclear (FT 2014), oil and gas (The Telegraph 2018) and aerospace (BAES 2020). While the UK has world-class additive manufacturing expertise in research, design and manufacturing many UK SMEs lack the awareness, resources or confidence to apply it as part of their manufacturing toolkit (UKAMSG 2017).

10.6 UK INDUSTRY 4.0 OPPORTUNITIES FOR SMEs

The UK Industry 4.0 landscape and drivers discussed earlier in this chapter are linked to a myriad of opportunities for SMEs. The role of UKRI in pushing industry's R&D investment towards acceptable risk levels was highlighted. This represents significant opportunities for SMEs as, under certain funding schemes, they would be eligible to higher recovery rates than larger businesses. Another opportunity that was alluded to earlier is the ongoing need for upgrades of high-tech research infrastructure and facilities, which also presents a platform for SMEs to secure contracts to conduct some of these upgrades, be it via tendering processes or bidding for innovation-focused calls by UKRI.

Opportunities were highlighted in relation to the net zero economy, and it was argued that the fact that 99 percent of UK firms are SMEs would inevitably mean they have a significant role in this national transformation. For example, it is noteworthy that the UK government had launched a taskforce to support a drive for 2 million green jobs by 2030 (BEIS Press Release 2020 (2)). The Industrial Decarbonisation Challenge (part of the ISCF) aims to reduce the carbon footprint of heavy and energy intensive industries and focusses on six large industrial clusters that sustain 1.5 million jobs and export up to £320 billion a year, but release around 40 million tonnes of carbon dioxide per year. It will develop at least one low-carbon industrial cluster by 2030, and the world's first net zero carbon industrial cluster by 2040 (BEIS Policy Paper 2021).

Upskilling (or re-skilling in the broader sense) has been discussed on several occasions in this chapter. A recent study by McKinsey & Co. (2020) suggested that around 30 percent of workforce reskilling cases in UK SMEs would result in a net benefit. Interestingly, the same study also suggested that, in the minority of 10 percent of cases across SMEs where reskilling was needed but employers did not profit from reskilling their workers, employers can still enjoy benefits if reskilling costs are complemented by state support. Therefore, government incentives for lifelong training and the push towards modularization of higher education (highlighted in the Drivers section) are expected to generate opportunities for SMEs not only to reduce the cost of upskilling their workers but also to proactively engage in providing the training and transfer of knowledge and know-how in partnership with Further and Higher Education institutions.

Additionally, a number of outstanding pillars of Industry 4.0 (e.g., cyber-security, AI, CFD, automation and robotics, and additive manufacturing) were highlighted as exemplary drivers that propel some significant innovation activity, unlocking further collaborative opportunities for SMEs.

Having reviewed a range of policies, drivers and challenges related to Industry 4.0 and relevant to SMEs in the UK, it is worth sampling a number of sectors of growing importance. Although not exclusively, they represent potential hotspots of opportunities for investment, collaboration, co-creation, and growth, particularly for SMEs:

1. Food and drink: the growing labor shortages in the UK's agricultural sector (DEFRA 2016) were compounded by COVID-19, which also exposed its over-reliance on vulnerable, international supply chains. The limits imposed on the number of certain items of foodstuffs that can be bought in UK supermarkets during the first half of 2020 drove home the need for upgrading UK farming, boosting its efficiency, and overcoming its reliance on manual labor. There is a need for more professional membership organizations such as Agri-TechE (www.agri-tech-e.co.uk) and IAgrE (www.iagre.org) that focus on facilitating step-changes in the UK's agriculture sector and bringing together farmers, researchers, innovators, government, and other stakeholders.

2. Foundation industries: comprising cement, glass, ceramics, paper, metals and bulk chemicals are worth £52 billion of the UK economy but generate 10 percent of the UK's CO2 emissions (Innovate UK 2020). Between 2000 and 2015, their share of UK GDP fell by 43 percent (Lawrence and Stirling 2016). The UK's ambitious net zero targets must not stifle the recovery, revival, or growth of these vital sectors. Rather, there is an opportunity to capitalize on the government's investment in foundation industries through UKRI (UKRI 2021 (4)) in line with its "Build Back Better" policy. Breaking the silos between various foundation industries, establishing a symbiotic relationship between their resource flows, and involving SMEs in the reuse and repurposing of the constituents of their processes can be achieved by greater adoption of Industry 4.0 technologies.
3. Textiles: the fast fashion industry is associated with a number of pressing challenges. Various observers point out that fashion supply chains are often marred by a lack of transparency that conceals unethical practices such as forced labor (Environmental Audit Committee 2019), excessive carbon footprint, and disproportionate amounts of waste from out-of-fashion textiles (WRAP 2017, 2019). In the UK, over 90 percent of the fashion industry in composed of SMEs (Alliance Project & N Brown 2017), and 52 percent of UK and US shoppers want the fashion industry to follow more sustainable practices and 29 percent of consumers would pay more for sustainably made versions of the same items (Moore 2019). Thus, there is an opportunity for SMEs to be a key part of the solution, by adopting Industry 4.0 technologies that allow higher traceability of supply chains, off-shoring of manufacturing processes, and better utilization of waste and by-products.
4. Construction: According to the UK's Construction Sector Deal (UK Government 2018), this industry employs around 3.1 million people, adding £138 billion to the economy – nearly 9 percent of GDP – and produces approximately 120 million tonnes of waste a year, nearly 60 percent of all UK waste. The Sector Deal acknowledges the need to accelerate the shift in towards digital processes, including, for example, the ISCF transforming construction challenge (UKRI 2021 (5)) which aims for: projects delivered 50 percent faster; whole life costs reduced by 33 percent; lifetime emissions halved; and productivity raised by 15 percent. Moreover, construction accounts for over 36 percent of occupational fatal injuries to workers in the UK (HSE 2020), many of which could be avoided through better evaluation using Industry 4.0 technologies such as automation, visualization, real-time monitoring, and digital twin through the application of Building Information Modelling (Craveiro et al., 2019).
5. International collaborations: Since the spending of a minimum percentage of GDP on overseas aid was enshrined in UK law in 2015 through Parliament's International Development (Official Development Assistance Target) Act, UKRI's contribution to this spend has been primarily through the Global Challenges Research Fund, whose aim is to ensure UK science takes the lead in addressing the problems faced by developing countries

while developing the ability to deliver cutting-edge research. Many of its main themes are underpinned by Industry 4.0 technologies, such as sustainable cities, and sustainable production and consumption of materials (UKRI 2021 (6)). Moreover, the UK's International Research and Innovation Strategy (UK Government 2019 (2)), sought to emphasize the UK's desire to establish more bilateral agreements with leading and emerging innovation powers on the world stage, attract world-class talent, and ensure its regulatory system enables foreign investment in opportunities stemming from the fourth Industrial Revolution. Additionally, post-Brexit Britain endeavors to offer wide-ranging opportunities for trade and innovation with a global reach. For example: (1) In addition to free trade provisions, the EU-UK Trade and Cooperation Agreement paved the way for its continued participation in Horizon Europe – the EU's research and innovation framework program running from 2021 to 2027 (www.ec.europa.eu/horizon-europe). (2) A growing list of trade agreements with non-EU countries (UK Government 2021 (1)). (3) Strategically, the government embraced the Integrated Review of Security, Defence, Development and Foreign Policy it commissioned in 2020/21 (UK Government 2021 (2)) and its recommendation to engage more deeply with the Indo-Pacific region, including the strengthening of supply chain resilience of critical goods and raw materials and deeper partnerships in science, technology, and data.

10.7 CONCLUSION

This chapter is a review of the UK's Industry 4.0 policy, science, technology, and industrial environment which SMEs can engage with in order for them to be part of the country's post-COVID-19, post-Brexit industrial and economic era. It attempts to outline three overarching aspects: (1) Research and innovation priorities and infrastructure have been the guiding beacon for Industry 4.0 activities in the UK, particularly those impacting SMEs; (2) The Made Smarter movement has defined the UK's brand of Industry 4.0, and is focused on scale-up, adoption and upskilling; (3) the focus on Net Zero Emissions and sustainability will dictate innovation and industrial policy throughout this decade.

UK SMEs can access an environment that nurtures innovation, although slow commercialization remains an obstacle. This was for many years exacerbated by the lack of leadership, vision, and branding of the UK's Industry 4.0 capabilities, although recently the landscape has been shifting to mitigate the shortfall in these aspects. In this chapter, we outlined with more detail three additional barriers that may be viewed as being more fundamental: low uptake from SMEs; user and regulator trust; and the upskilling challenge and shortage of trained graduates.

The evolving landscape – reinforced by policy initiatives such as the Industrial Strategy, Made Smarter and the Levelling-up Agenda – is helping nurture a variety of drivers for the adoption of Industry 4.0 technologies. In this chapter some drivers – paving the way for the increased use of Industry 4.0 – were highlighted: (1) Technological drivers such as the accelerating developments in advanced materials, post-COVID pharmaceuticals and the rapid increases in computational power, (2) Economic, such as the need for frictionless trade, ethical production and global collaboration, and (3) Social and environmental, including climate change, urban manufacturing and manufacturing symbiosis.

To different but certain extents, the barriers and drivers generate a myriad of opportunities for UK SMEs in the Industry 4.0 space, some of which were highlighted in this chapter. These include: (1) A hospitable environment for R&D investment, (2) Continuous need for upgrades and digitalization of high-tech and research infrastructure, (3) Outstanding pillars of Industry 4.0 where the UK has – or seeks to – become a world leader, including (but not limited to) cyber security, AI, computational fluid dynamics, automation & robotics, and additive manufacturing, (4) A national drive for green jobs, and (5) A range of opportunity-laden sectors that are increasingly pivotal to the emerging industrial scene in the UK and heavily reliant on SMEs, exemplified by: Food and drink, foundation industries, textiles, the construction industry, and the internationally facing R&D and capacity building collaborations.

Industry 4.0 is already part of the fabric of the UK's R&D activities, and Made Smarter has been instrumental in defining and articulating the UK's Industry 4.0 offering. The future narrative of Industry 4.0 in the UK will be largely influenced by the net zero agenda, in which SMEs and research institutions alike will have a plethora of collaborative opportunities to take advantage of.

10.8 ACKNOWLEDGEMENTS

The author thanks Dimitar Karamanchev Co-founder and Head of R&D and Strategic Partnerships at IndustriGen Ltd for many, valuable conversations about their work in the area of the adoption of Industry 4.0 by SMEs, and Scott Pepper from GAMBICA (the Trade Association for Instrumentation, Control, Automation and Laboratory Technology in the UK) for the valuable insight into the challenges and needs of UK SMEs. Special thanks to Prof. Paulo Bartolo, Nanyang Technological University and the University of Manchester, whose leadership and friendship were the reason why the author is in a position to contribute to this publication.

REFERENCES

Alliance Project & N Brown, 2017. Realising the Growth Potential of UK Fashion & Textile Manufacturing Report. National Textiles Growth Programme. Available from www.ltma.co.uk/wp-content/uploads/2017/05/The-Final-Alliance-Project-Report-Oct-2012-to-May-2017.pdf [Accessed 30 March 2021].

Augar, P., 2019. Review of Post-18 Education and Funding. Commissioned by the UK Government. Available from https://assets.publishing.service.gov.uk/government/uploads/system/uploads/attachment_data/file/805127/Review_of_post_18_education_and_funding.pdf [Accessed 30 March 2021].

BAE Systems (BAES), 2020. Developing 3D Printing Capability for the Defence and Aerospace Sector. Available from www.baesystems.com/en/article/developing-3d-printing-capability-for-the-defence-and-aerospace-sector [Accessed 30March 2021].

Beatty, C., and Fothergill, S., 2016. Jobs, Welfare and Austerity. Centre for Regional Economic and Social Research at Sheffield Hallam University. Available from https://

www4.shu.ac.uk/research/cresr/sites/shu.ac.uk/files/cresr30th-jobs-welfare-austerity.pdf [Accessed 30 March 2021].
BEIS Press Release, 2020 (1). Available from www.gov.uk/government/news/new-productivity-institute-part-of-37m-investment-to-boost-uk-wage-growth-and-living-standards [Accessed 30 March 2021].
BEIS Press Release, 2020 (2). Available from www.gov.uk/government/news/uk-government-launches-taskforce-to-support-drive-for-2-million-green-jobs-by-2030 [Accessed 30 March 2021].
BEIS and RSM PACEC Ltd, 2018. Research into Issues Around the Commercialisation of University IP.
BEIS, UK, 2020. Business Population Estimates for the UK and Regions. Available from www.gov.uk/government/statistics/business-population-estimates-2020 [Accessed 30 March 2021].
BEIS, UK, Policy Paper, Updated 2021. Available from www.gov.uk/government/publications/industrial-strategy-the-grand-challenges/missions [Accessed 30 March 2021].
Business, Energy & Industrial Strategy (BEIS) Department, UK, 2017. Available from gov.uk/government/publications/industrial-strategy-building-a-britain-fit-for-the-future [Accessed 30 March 2021].
Cambridge Industrial Innovation Policy (CIIP), 2020. The Role of Industrial Digitalisation in Post-Covid-19 Manufacturing Recovery, Diversification and Resilience.
Certara Ltd, 2016. The Benefits of Modelling and Simulation in Drug Development. Available from www.certara.com/app/uploads/2020/06/WP_BenefitsOfBiosimulation-1-1.pdf [Accessed 30 March 2021].
Comrie, M., 2018. Feasibility Study: Growing Industry 4.0 Capability in SMEs. Available from https://admin.ktn-uk.co.uk/app/uploads/2018/12/Growing-The-Industry-4.0-Capability-of-SMEs_mj-v3.pdf [Accessed 30 March 2021].
Coop, 2019. Twenty Years of Ethical Consumption. Available from https://assets.ctfassets.net/5ywmq66472jr/5hkc6bA1y2eNRGsHJzyvX2/14449115fafac1c02cf4f9fd5a52b13b/Twenty_Years_of_Ethical_Consumerism_2019.pdf [Accessed 30 March 2021].
Craveiro, F., Duarte, J. P., Bartolo, H., and Bartolo, P. J., 2019. Additive Manufacturing as an Enabling Technology for Digital Construction: A Perspective on Construction 4.0. *Automation in Construction* 103, 251–267.
Deloitte, 2021. Manchester Crane Survey. Available from https://www2.deloitte.com/uk/en/pages/real-estate/articles/manchester-crane-survey.html [Accessed 30 March 2021].
Department for Education (DfE), UK, 2021. Interim Conclusion of the Review of Post18 Education and Funding. Available from https://assets.publishing.service.gov.uk/government/uploads/system/uploads/attachment_data/file/953332/Interim_Conclusion_of_Review_of_Post-18_Education_and_Funding.pdf [Accessed 30 March 2021].
Department for Environment, Food & Rural Affairs (DEFRA), 2016. Agricultural Labour in England and the UK, Farm Structure Survey. Available from https://assets.publishing.service.gov.uk/government/uploads/system/uploads/attachment_data/file/771494/FSS2013-labour-statsnotice-17jan19.pdf [Accessed 30 March 2021].
Diffey, C., 2019.2018 Funding for UK AI Sector Hit a Record £998m. TechRound. Available from https://techround.co.uk/news/2018-funding-for-uk-ai-1–3bn/ [Accessed 30 March 2021].
Dowling, A., 2015. Review of Business-University Research Collaborations. Availabel from www.raeng.org.uk/publications/reports/the-dowling-review-of-business-university-research [Accessed 30 March 2021].
During, L., 2019. The UK's AI Landscape: A Tale of AI, Manufacturing and the Avengers. Make UK Blogs. Available from www.makeuk.org/insights/blogs/the-uks-ai-landscape-a-tale-of-ai-manufacturing-and-the-avengers [Accessed 30 March 2021].
EEF, 2016. An Up-Skill Battle: EEF Skills Report 2016. Available from www.makeuk.org/~/media/c669e0f103654add9fb2224fb53bc8b3.pdf [Accessed 30 March 2021].

Engineering UK, 2020. Educational Pathways into Engineering. Available from www.engineeringuk.com/media/196594/engineering-uk-report-2020.pdf [Accessed 30 March 2021].

Environmental Audit Committee, UK Parliament, 2019. Fixing Fashion: Clothing Consumption & Sustainability. Available from https://publications.parliament.uk/pa/cm201719/cmselect/cmenvaud/1952/report-summary.html [Accessed 30 March 2021].

Financial Times (FT), 2014. Sellafield Hopes to Allay Cost Fears with 3D Printing. Available from www.ft.com/content/42c1d086-d779-11e3-a47c-00144feabdc0 [Accessed 30 March 2021].

Garton, G., and Barton, C., 2020. Briefing Paper on Green Belt, House of Commons Library.

Greater Manchester Independent Prosperity Review (GMIPR), 2019. Available from www.greatermanchester-ca.gov.uk/what-we-do/economy/greater-manchester-independent-prosperity-review/ [Accessed 30 March 2021].

Health and Safety Executive (HSE), 2020. Workplace Fatal Injuries in Great Britain. Available from www.hse.gov.uk/statistics/pdf/fatalinjuries.pdf [Accessed 30 March 2021].

House of Commons (HoS) Business, Energy and Industrial Strategy (BEIS) Committee, 2019. Automation and the Future of Work. Available from https://publications.parliament.uk/pa/cm201719/cmselect/cmbeis/1093/1093.pdf [Accessed 30 March 2021].

Hughes, A., and Atkinson, P., 2018. England's North-South Divide Is History – but the Nation's Rifts Are Deepening. *The Conversation*. Available from https://theconversation.com/englands-north-south-divide-is-history-but-the-nations-rifts-are-deepening-99044 [Accessed 30 March 2021].

Innovate UK, 2020. Available from www.gov.uk/government/news/foundation-industries-building-a-resilient-recovery-apply-for-funding [Accessed 30 March 2021].

Innovate UK, 2021. Available from www.gov.uk/government/organisations/innovate-uk [Accessed 30 March 2021].

Institution of Engineering and Technology (IET), 2015. Skills and Demand in Industry. Available from www.theiet.org/media/4384/iet-skills-survey-2015.pdf [Accessed 30 March 2021].

Jarrahi, M. H., 2018. Artificial Intelligence and the Future of Work: Human-AI Symbiosis in Organizational Decision-making. *Business Horizons* 61, 4, 577–586.

Karlsson, L., 2018. Smart Border 2.0 – Avoiding a Hard Border on the Island of Ireland for Customs Control and the Free Movement of Persons. EU Policy Department for Citizens' Rights and Constitutional Affairs.

Lawrence, M., and Stirling, A., 2016. Strong Foundation Industries: How Improving Conditions for Core Material Producers Could Boost UK Manufacturing. Institute for Public Policy Research. Available from www.ippr.org/files/publications/pdf/strong-foundation-industries_March2016.pdf?noredirect=1 [Accessed 30 March 2021].

Local Government Association Website. Available from www.local.gov.uk/topics/devolution/devolution-online-hub/devolution-explained/devolution-deals [Accessed 30 March 2021].

Made Smarter Review, 2017. Available from www.gov.uk/government/publications/made-smarter-review [Accessed 30 March 2021].

Make UK, 2020. UK Manufacturing Facts 2020/21. Available from makeuk.org [Accessed 30 March 2021].

Manchester Zero Carbon Framework 2020–38 (Draft). Available from: www.manchesterclimate.com/framework-2020-2038 [Accessed 30 March 2021].

McKinsey & Company, 2020. The Economic Case for Reskilling in the UK: How Employers Can Thrive by Boosting Workers' Skills. Available from www.mckinsey.com/business-functions/organization/our-insights/the-economic-case-for-reskilling-in-the-uk-how-employers-can-thrive-by-boosting-workers-skills# [Accessed 30 March 2021].

Moore, K., 2019. Report Shows Customers Want Responsible Fashion, But Don't Want To Pay For It. What Should Brands Do? Forbes. Available from https://www.forbes.com/sites/

kaleighmoore/2019/06/05/report-shows-customers-want-responsible-fashion-but-dont-want-to-pay-for-it/ [Accessed 30 March 2021]

Northern Powerhouse, UK Government, 2016. Northern Powerhouse Strategy. Available from https://assets.publishing.service.gov.uk/government/uploads/system/uploads/attachment_data/file/571562/NPH_strategy_web.pdf [Accessed 30 March 2021].

Rolls Royce, 2021. Available from www.rolls-royce.com/about/our-research/research-and-university.aspx [Accessed 30 March 2021].

Sheffield Political Economy Research Institute (SPERI), 2015. Public and Private Sector Employment Across the UK Since the Financial Crisis. Available from http://speri.dept.shef.ac.uk/wp-content/uploads/2018/11/Brief10-public-sector-employment-across-UK-since-financial-crisis.pdf [Accessed 30 March 2021].

Siemens, 2019. Available from https://news.siemens.co.uk/news/siemens-launches-industry-4-0-curriculum-for-universities [Accessed 30 March 2021].

SWMAS, 2020. Manufacturing Barometer, National Report for Q3 2020/21 (Special Focus: Trading Beyond Brexit). Available from www.swmas.co.uk/knowledge/national-2020-q3. [Accessed 30 March 2021].

The Telegraph, 2018. 3D Printing Will 'Transform' the Oil Industry, Says BP. Available from www.telegraph.co.uk/technology/2018/12/29/3d-printing-will-transform-oil-industry-says-bp/ [Accessed 30 March 2021].

Treasury, UK, 2020 (1). Budget. Available from www.gov.uk/government/publications/budget-2020-documents/budget-2020 [Accessed 30 March 2021].

Treasury, UK, 2020 (2). National Infrastructure Strategy. Available from https://assets.publishing.service.gov.uk/government/uploads/system/uploads/attachment_data/file/938539/NIS_Report_Web_Accessible.pdf [Accessed 30 March 2021].

Treasury, UK, 2021. Build Back Better: Our Plan for Growth. Available from https://assets.publishing.service.gov.uk/government/uploads/system/uploads/attachment_data/file/968403/PfG_Final_Web_Accessible_Version.pdf [Accessed 30 March 2021]

UK Additive Manufacturing Steering Group (UKAMSG), 2017. Additive Manufacturing UK National Strategy 2018–25.

UK Commission for Employment and Skills (UKCES), 2015. Sector Insights: Skills and Performance Challenges in the Advanced Manufacturing Sector. Available from via www.gov.uk/government/publications/sector-insights-skills-and-performance-challenges-in-the-advanced-manufacturing-sector [Accessed 30 March 2021].

UK Government, 2018. Industrial Strategy Construction Sector Deal. Available from https://assets.publishing.service.gov.uk/government/uploads/system/uploads/attachment_data/file/731871/construction-sector-deal-print-single.pdf [Accessed 30 March 2021].

UK Government, 2019 (1). Greater Manchester Local Industrial Strategy (LIS). Available from www.investinmanchester.com/dbimgs/190612%20Final%20GM%20Local%20Industrial%20Strategy.pdf [Accessed 30 March 2021].

UK Government, 2019 (2). International Research and Innovation Strategy. Available from https://assets.publishing.service.gov.uk/government/uploads/system/uploads/attachment_data/file/801513/International-research-innovation-strategy-single-page.pdf [Accessed 30 March 2021].

UK Government, 2020 (1). The Ten Point Plan for a Green Industrial Revolution. Available from https://assets.publishing.service.gov.uk/government/uploads/system/uploads/attachment_data/file/936567/10_POINT_PLAN_BOOKLET.pdf [Accessed 30 March 2021].

UK Government, 2020 (2). UK Research and Development Roadmap. Available from www.gov.uk/government/publications/uk-research-and-development-roadmap. [Accessed 30 March 2021].

UK Government, 2020 (3). Government Response to the BEIS Select Committee's Twenty-Third Report. Available from https://publications.parliament.uk/pa/cm5801/cmselect/cmbeis/240/24002.htm [Accessed 30 March 2021].

UK Government, 2021 (1). UK Trade Agreements with Non-EU Countries. Available from www.gov.uk/guidance/uk-trade-agreements-with-non-eu-countries#trade-agreements [Accessed 30 March 2021].

UK Government, 2021 (2). Global Britain in a Competitive Age: The Integrated Revie of Security, Defence, Development and Foreign Policy. Available from https://assets.publishing.service.gov.uk/government/uploads/system/uploads/attachment_data/file/975077/Global_Britain_in_a_Competitive_Age-_the_Integrated_Review_of_Security__Defence__Development_and_Foreign_Policy.pdf [Accessed 30 March 2021].

UK Research and Innovation (UKRI), 2020 (1). The UK's Research and Innovation Infrastructure: Landscape Analysis. Available from www.ukri.org/wp-content/uploads/2020/10/UKRI-201020-LandscapeAnalysis-FINAL.pdf [Accessed 30 March 2021].

UKRI, 2020 (2). The UK's Research and Innovation Infrastructure: Opportunities to Grow Our Capability. Available from www.ukri.org/wp-content/uploads/2020/10/UKRI-201020-UKinfrastructure-opportunities-to-grow-our-capacity-FINAL.pdf [Accessed 30 March 2021].

UKRI, 2021 (1). Available from www.ukri.org/our-work/our-main-funds/industrial-strategy-challenge-fund/ [Accessed 30 March 2021].

UKRI, 2021 (2). Available from www.ukri.org/our-work/our-main-funds/industrial-strategy-challenge-fund/future-of-mobility/robots-for-a-safer-world-challenge/ [Accessed 30 March 2021].

UKRI, 2021 (3). Transforming Our World with AI'. Available from www.ukri.org/wp-content/uploads/2021/02/UKRI-120221-TransformingOurWorldWithAI.pdf. [Accessed 30 March 2021].

UKRI, 2021 (4). Transforming Foundation Industries Challenge. Available from www.ukri.org/our-work/our-main-funds/industrial-strategy-challenge-fund/clean-growth/transforming-foundation-industries-challenge/ [Accessed 30 March 2021].

UKRI, 2021 (5). Transforming Construction Challenge. Available form www.ukri.org/our-work/our-main-funds/industrial-strategy-challenge-fund/clean-growth/transforming-construction-challenge/ [Accesse 30 March 2021].

UKRI, 2021 (6). Global Challenges Research Fund. Available from www.ukri.org/our-work/collaborating-internationally/global-challenges-research-fund/ [Accessed 30 March 2021].

University of Manchester (UoM), 2016. Greater Manchester and Cheshire East: A Science and Innovation Audit Report. Available from https://documents.manchester.ac.uk/display.aspx?DocID=30337 [Accessed 30 March 2021].

Waste & Resources Action Programme (WRAP), 2017. Valuing Our Clothes: The Cost of UK Fashion. Available from https://wrap.org.uk/resources/report/valuing-our-clothes-cost-uk-fashion [Accessed 30 March 2021].

Waste & Resources Action Programme (WRAP), 2019. Textiles: Market Situation Report. Available from https://wrap.org.uk/resources/market-situation-reports/textiles-2019 [Accessed 30 March 2021].

Wohlers Report, 2021. 3D Printing and Additive Manufacturing State of the Industry. ISBN 978-0-9913332-7-1.

World Bank Data, 2020. Available from worldbank.org [Accessed 30 March 2021].

World Health Organisation (WHO), 2021. Inside the Mammoth Undertaking of Global Vaccine Distribution. Interview with Katherine O'Brien, Director of the Department of Immunization, Vaccines and Biologicals. Available from www.who.int/news-room/feature-stories/detail/inside-the-mammoth-undertaking-of-global-vaccine-distribution [Accessed 30 March 2021].

11 The Path Towards Industry 4.0 for Brazilian SMEs

Gilson Adamczuk Oliveira, Marcelo Gonçalves Trentin, Dalmarino Setti, and Jose Donizetti de Lima

CONTENTS

11.1 INTRODUCTION

Industry 4.0 is the incorporation of digitalization into industrial activity, integrating physical and virtual technologies. The main technologies include big data, the Internet of Things, advanced robotics, cloud computing, additive manufacturing, artificial intelligence, machine-to-machine connection systems (M2M), sensors, actuators, and advanced production management software (CNI 2017). Despite being one of the more important topics for industry, when Industry 4.0 is considered in terms of SMEs, the scarcity of academic studies worldwide is remarkable.

SMEs are fundamental in both high- and low-income economies worldwide. In the Brazilian emerging economy, SMEs play a crucial role in GDP growth and employment generation (EdinburghGroup 2014). Brazil has a predominance of SMEs in the national economy and that operate outside high-tech sectors. They have capital restrictions, outdated facilities, and low performance in the international market. These organizations have regional difficulties with inadequate infrastructures and need more integration with research, development and innovation institutions, and business networks participation.

DOI: 10.1201/9781003165880-11

On the other hand, there are Brazilian companies that have a leading position in the international scenario (agricultural, mining, banking, and automobile sectors) (BRASIL 2017). Approximately 99 percent of enterprises in Brazil are SMEs, which account for about 52 percent of jobs generated by companies in the country. However, they produce only 25 percent of the GDP in the business sector and 27 percent of the Brazilian GDP (Schaefer et al. 2020).

SMEs are, on average, less productive than large firms. In developing countries, they are 70 percent less efficient than large companies. SMEs are incapable of taking advantage of scale economies. They have difficulties getting access to credit lines, and there is a lack of appropriate skills and a high level of informality (WTO 2016). Consequently, SMEs struggle to foment projects related to Industry 4.0. Scarcity of resources urges us to identify and prioritize opportunities and make appropriate decisions. Therefore, substantial barriers need to be overcome by SMEs for Industry 4.0 implementation (SEBRAE 2014).

Motivated by this scenario, this chapter investigates the barriers, drivers, and opportunities in Brazil with regard to digitalization. The aim is to present the Brazilian perspective under the new manufacturing paradigm. In addition to this brief introduction, this chapter is organized as follows. The next section shows the country background and digitalization policy in Brazil. After the methodology strategy, the chapter is concluded with the barriers, drivers, and opportunities concerning Industry 4.0 application in national SMEs.

11.2 COUNTRY BACKGROUND

Brazil's manufacturing sector occupies 40th position in the Competitive Industrial Performance Index 2020 and accounts for 9.2 percent of national GDP (UNIDO 2020), almost a third of its contribution of over 30 percent in the 1980s. This reduction is due to changes in the productive structure of the country and new business models brought about by technological disruption. According to the Readiness for the Future of Production Report 2018, Brazil is 41st position in Structure of Production (5.22 score) and the 47th position in the Drivers of Production (5.08 score) (The World Bank 2020).

Brazil is positioned in the nascent category among 100 countries. Leading segments of the Brazilian manufacturing sector are food and beverages 22.6 percent, chemicals and chemical products 13.8 percent, coke, refined petroleum products, nuclear fuel 8.6 percent, motor vehicles, trailers, semi-trailers 7.3 percent, and machinery and equipment 6.9 percent. In terms of the structure of production, the manufacturing sector has the following composition: resource-based 48.1 percent, low-tech 8.7 percent, medium-tech 37.0 percent, and high-tech 6.2 percent (UNIDO 2020).

Digitalization in the Brazilian Industry has prioritized the production processes, i.e., increasing efficiency and productivity. Considering only national companies that adopt digital technologies, 73 percent adopt at least one of such technologies into their production processes; 47 percent are related to developing the production chain and 33 percent to new products and businesses (CNI 2016a).

Brazil in 2008 received 1,202 patent applications for inventions related to Industry 4.0 technologies, which represented 5 percent of the total of 23,170 applications made that year. A decade later, Brazil has 14,634 "I4.0 patents," representing 57 percent of the total of 25,658 applications in 2017. This exploratory study on patents investigated the technologies related to Industry 4.0, which were analyzed and grouped into central technologies, enabling technologies and application sectors. Central technologies involve hardware, software, and basic connectivity systems, transforming a product into an intelligent device connected to the internet. Enabling technologies of Industry 4.0 are technologies built on and complementary to central technologies. They comprise data analysis systems, user interface, physical and simulated 3D systems, artificial intelligence, location systems, power systems, and security systems. Application sectors cover Industry 4.0 technologies for end-users: personal, home, automotive, industry, and cities (CNI 2020).

11.3 DIGITALIZATION POLICY IN THE COUNTRY

In Brazil, actions and policies to encourage technological development and innovation have been taking place since the early 2000s, but at a slow pace. Government laws and decrees updated national legislation and directed incentives to stimulate this development. The first concrete federal programs started in 2017 (BRASIL 2018b)(ABGI BrasiI 2018).

In 2018, the National Digital Transformation System became official, and with it, the governance structure for the implementation of the Brazilian Digital Transformation Strategy named "E-Digital" (BRASIL 2018a, 2019). In the country, the Ministry of Science, Technological Innovations and Communications (MCTIC) is responsible for formulating digital transformation policies and the national Internet of Things plan (IoT.BR). This body articulates with civil society, the scientific community, the government, and the productive sector. E-Digital was promulgated in 2018 (BRASIL 2018b).

IoT.BR seeks to accelerate the dissemination of technologies and the extensive use of the Internet of Things and to assess its impact on all economic sectors. The plan is considered one of the bases of the digitalization process of the Brazilian economy. Among the various actions proposed, it was highlighted how to promote the adoption of the IoT in SMEs both in the selection and implementation of adequate solutions (BRASIL 2018c).

E-Digital was founded based on two groups of axes: (1) enabling axes (infrastructure, access to technologies, protection and security, education and training, internationalization in the digital economy) and (2) digital transformation axes (of the economy, citizenship and government). Currently, the country is in the 80th position of the "Global Competitiveness Index – UN." The strategy seeks to improve this position in five years (BRASIL 2018b). In addition to the general context, E-Digital includes some specific considerations regarding SMEs (BRASIL 2018b), such as:

- Demand for incentives and support for the internationalization process of SMEs: promotion and training actions, including the managers, mainly for international electronic commerce.

- Recognition of the relevant role of SMEs in generating jobs and generally higher regard for SMEs in the digital economy.
- Establishment of agreements and partnerships with international marketplaces supporting the export of Brazilian products and services on the internet by Brazilian SMEs. "Apex Brazil" and "e-Xport Brazil" are good examples of such initiatives.
- Promotion of strategic actions, joint work between authorities, and regulatory frameworks related to data to facilitate the entry of SMEs in global markets.
- Recognition that application platforms are favorable environments for proposing products and services, a space for innovation and development for SMEs.
- Logistics programs and services and access to financing funds to encourage SMEs to operate in electronic commerce.

The MCTIC, based on the "National Science, Technology and Innovation Strategy for the period 2016 to 2022" (ENCTI 2016/2022), considering studies by the National Confederation of Industry (CNI) and other works, proposed the "CT&I Plan for Advanced Manufacturing in the Brazil" (ProFuturo). The plan was built based on the governmental, academic, business sector, and international cooperation with the IDB, UNESCO, EU, and reference countries. The strategy mentions the importance of engagement. It has an immediate impact on organizations and economies and considers technology, human resources, production chains, infrastructure, and regulation. These areas are related to the dimensions of government rules, academia, and companies.

One of the goals outlined by ProFuturo focuses on SMEs: "Availability of instruments, means and conditions for access and insertion of SMEs in the advanced manufacturing ecosystem." This goal has actions related to legislation, the creation of instruments to induce, promote access to advanced manufacturing, including the adaptation of current production structures (BRASIL 2017).

Authorities and society created the "Digital Transformation Observatory" as one of the actions to guarantee the participation and monitoring of public policies on the subject, a partnership involving civil society, the business sector, and the federal government. Other digitalization promoters are the Telecommunications Research and Development Center (CPqD) and the Competitive Brazil Movement (MBC). The first is a private organization and one of the principal innovation centers in information and communication technology in Latin America. All these bodies, with governance and public management, have been stimulating the debate and the initiatives for promoting economic and sustainable growth. The observatory helps disseminate and promote E-Digital, the "CT&I Plan for Advanced Manufacturing in Brazil" (ProFuturo), and the IoT.BR (BRASIL, CPqD, and MBC 2020).

In 2018 the MCTIC and the National Service for National Learning (SENAI) developed the project "Mapping 4.0" to identify and map the initiatives carried out in the country. This information makes up the government's database and supports decisions, policies, and actions related to technology and innovation in Industry 4.0 (BRASIL and SENAI 2020).

11.4 METHODOLOGY

This chapter explores the *barriers*, *drivers*, and *opportunities* for Industry 4.0 implementation in Brazilian SMEs. It is a conceptual research approach that identifies the variables of interest and the expected relationships (Bickman and Rog 2009). We opted for secondary data sources, due to the expected incipience of the digitalization in national SMEs and the difficulties of collecting data during the COVID-19 pandemic in Brazil.

Usual secondary data sources for social science include censuses, information collected by government agencies, organizational records, and data from previous research. In this context, we highlight data from the Brazilian Ministry of Economy (ME), the National Confederation of Industry (CNI), the federations of industries in the states (FIs), the National Service of Industrial Training (SENAI), and the Brazilian Service of Support for Micro and Small Enterprises (SEBRAE).

We gave priority to Portuguese texts, understanding it as reasonable to give a picture of the current Brazilian scenario to a global audience regarding Industry 4.0 and SMEs. We restricted the search to the period 2014–2020, considering the speed of the technological changes. Also, because the term *Industry 4.0* was coined in 2011, it is reasonable to collect data a few years later, considering a natural latency until related information should be published. Finally, data from national sources are compared with international references, providing external validity to the findings.

ME, CNI, FIs, and SENAI provide capillarity, i.e., information about the whole country. The ME is leading the actions and policies to support the digitalization in the Brazilian Industry. Therefore, it is a good source for examining the *opportunities* and the *drivers* regarding Industry 4.0. CNI, FIs, and SENAI are a rich source for a general perspective of the *barriers* to implement Industry 4.0 in Brazil. Finally, SEBRAE is a private non-profit entity that focuses on training and development. It is an organization created to support small businesses.

The document analysis involved superficial examination, reading (thorough examination), and text interpretation. It is an iterative process that combines elements of content analysis and thematic analysis. Content analysis is the process of organizing information into categories related to the central research questions (Bowen 2015). Based on the critical analysis of information sources, suggestions, tools, and actors are presented that may help in actions to overcome the challenges to the implementation of Industry 4.0 in Brazilian SMEs.

In order to achieve the goals of a digitalization strategy, the following features of Industry 4.0 should be considered: horizontal interconnection through value networks, vertical interconnection and networked manufacturing systems, and end-to-end digital integration of engineering across the whole value chain (Kagermann, Wahlster, and Helbig 2013). We added the importance of SMEs getting closer to educational and research institutions (Lehmann 2019), through knowledge transfer and training, accelerating the implementation process. Figure 11.1 illustrates the actors involved, on which the remaining sections of this chapter are based. Considering a balanced approach with both integrations improving, an optimal "gradient vector" is guided by Industry 4.0 adoption drivers. Finally, the barriers push against this process.

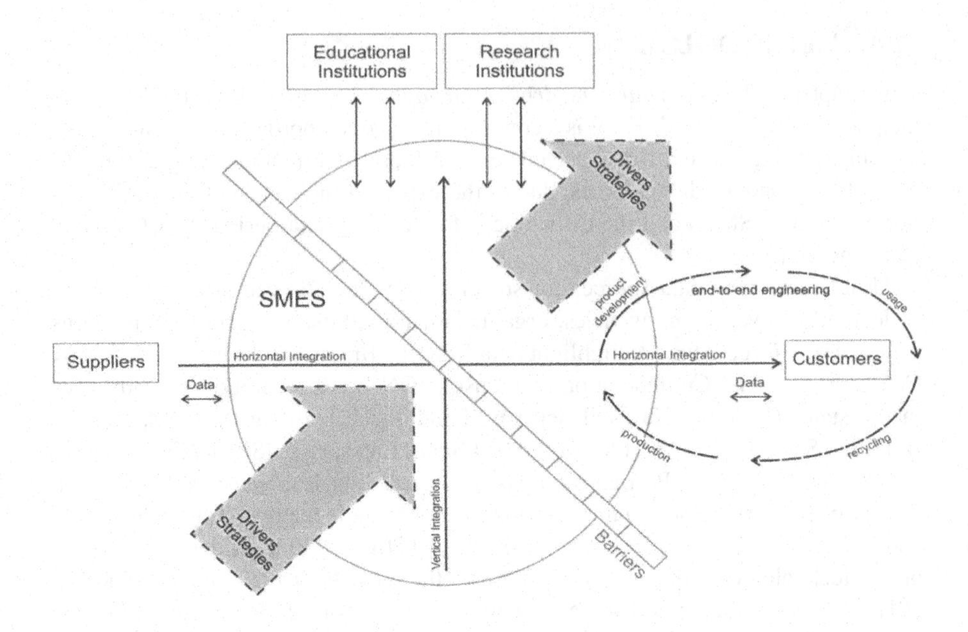

FIGURE 11.1 Actors involved in the digitalization strategy for SMEs

11.5 INDUSTRY 4.0 BARRIERS

Aligned with (Horváth and Szabó 2019), we organized the implementation obstacles into six dimensions: human resources, financial resources, management reality, organizational factors, technological factors, and regulation. Table 11.1 shows the main findings.

Even considering the limited number of documents on the Brazilian context, human resources, financial resources, organizational factors, and technological factors are similar both in a national and worldwide perspective. The diversity found in Brazil, a continental country (8,514,887 km^2), explains this convergence. The global view on barriers into these dimensions is more detailed and gives some important clues for the future of Industry 4.0 in Brazil. For example, "disruption to existing jobs" and "resistance by employees and middle management" are not yet in the spotlight in Brazil. These barriers are not externalized frequently yet. This is partially explained because digitalization in Brazil is still incipient compared with that of developed countries already committed to digitalizing their economies. We believe this technological gap also explains the non-externalization of barriers in Brazil on "management reality and regulatory." In a nutshell, digitalization in Brazil is still at a low maturity level, especially regarding SMEs. These two barriers' dimensions appear to be more urgent for organizations that are also more digitalized than Brazilian companies.

TABLE 11.1
Barriers for Digitalization in SMES

Brazil	Worldwide
Human Resources:	
	Lack of appropriate competences and skilled workforce (6)
Human resources not prepared for the digitalization (1) (5)	Longer learning time (training of staff) (6)
	Disruption to existing jobs (7) (8)
Lack of IT staff (3) (5)	Need for Enhanced Skills (7)
	Lack of digital skills (8)
Financial resources:	
	Lack of financial resources (6)
	Return and profitability (6)
High implementation costs (2) (3) (4) and (5)	Shortcomings in tendering systems (6)
	Long evaluation period for tenders (6)
Lack of clarity of return on investment (3) and (5)	High Implementation Cost (7) (8)
	Lack of clarity regarding the economic benefit (8)
Management reality:	
	Lack of a leader with appropriate skills, competencies, and experience (6)
Barriers not externalized in this dimension	Lack of conscious planning: defining goals, steps and needed resources (6)
	Lack of knowledge management systems (7)
	Ineffective change management (8)
Organizational factors:	
	Inadequate organizational structure and process organization (6)
	Contradictory interests in different organizational units (6)
Corporate structure and culture (3)	Resistance by employees and middle management (6)
	Organizational and process changes (7)
Lack of digitalization strategy (5)	Resistance to change (8)
	Lack of a digital strategy combined with resource scarcity (8)
	Lack of internal digital culture and training (8)
	Challenge in value-chain integration (8)

(Continued)

TABLE 11.1 (continued)
Barriers for Digitalization in SMES

Brazil	Worldwide
Technological factors:	
Difficulties in integrating new technologies and software (2) (3)	Lack of a unified communication protocol (6)
Information security risks (Cyber threats) (2) (3) (4) and (5)	Lack of back-end systems for integration (6)
Inappropriate IT infrastructure (3) and (5)	Lack of willingness to cooperate (at the supply chain level) (6)
	Lack of standards incl. technology and processes (6)
	Lack of proper, common thinking (6)
	Unsafe data storage systems (6)
	The need for large amounts of storage capacity (6)
	Lack of clear comprehension about IoT benefits (7)
	Lack of standards and reference Architecture (7)
	Lack of Internet coverage and IT facilities (7)
	Security and Privacy Issues (7)
	Seamless integration and compatibility issues (7)
	Risk of security breaches (8)
	The low maturity level of the desired technology (8)
	Lack of infrastructure (8)
	Challenges in ensuring data quality (8)
Regulation:	
	Regulatory Compliance issues (7)
Barriers not externalized in this dimension	Legal and Contractual Uncertainty (7)
	Lack of standards, regulations, and forms of certification (8)

Source: (1) BRASIL 2017, (2) BSA 2018, (3) CNI 2016b, (4) Lehmann 2019, (5) Stefanuto 2019, (6) Horváth and Szabó 2019, (7) Kamble, Gunasekaran, and Sharma 2018, and (8) Raj et al. 2020

11.6 INDUSTRY 4.0 DRIVERS

It is pertinent to present factors that may inspire companies to move towards the concept of Industry 4.0 (Horváth and Szabó 2019). We grouped together the drivers for barriers in human resources and management reality, as we understand they tackle these two dimensions. These drivers focus on education and training for the future workforce and retraining all actors currently involved in the digitalization efforts. Table 11.2 shows the main findings.

Although it is difficult to prioritize these drivers in this exploratory study, we can highlight the two most recurring of those cited in our small sample. Firstly, the investment in new educational models and training programs enforces the importance of SMEs getting closer to educational and research institutions

TABLE 11.2
Drivers for Digitalization in SMES

Brazil	Worldwide
Human Resources and Management reality:	
Investment in new educational models and training programs (1) (2) (4) (5) (6)	Allocating workforce to other areas (higher added value) (7)
Reformulation of undergraduate courses in the areas of engineering, administration and among others, to adapt the new needs of the technologies 4.0 (2)	Reducing human work (7)
Promotion of technical fairs, seminars and congresses related to digitalization (2)	Demand for greater control (from top management) (7)
Interaction with universities and research institutes (4) (5) (6)	Continuous monitoring of company performance (7)
Hiring professionals specialized in the new Industry 4.0 technologies (4)	Development of digital skills in non-production functions (8)
Financial resources:	
Establishment of specific credit lines (1) (2) (5) (6)	Reducing costs, e.g., human resources, inventory management and operating costs (7)
Low-cost solutions to start using technologies of Industry 4.0 (3) (5)	Public investments in terms of internet connection and infrastructure (9)
Incremental implementation as a digitalization strategy appropriate for national SMEs (4)	
Organizational factors:	
Collaboration between the private sector and with governments of other countries to address issues related to data transfer and security (1) (5)	Reducing the error rate Improving lead times (compliance with market conditions) (7)
Collaborative partnerships, planning and strategy towards the digitalization (4) (5)	Improving efficiency (7)
Development of Industry 4.0 in other countries increase competitive pressure on Brazilian companies towards digitalization (2)	Ensuring reliable operation (e.g., less downtime) (7)
Leading companies being digitalized encourage the others to shift to the new paradigm (2)	Change in culture towards new technologies (7), (10)
	Development of a department-arching understanding and plan for Industry 4.0 (11)
Technological factors:	
Development of digital infrastructure for companies (1) (2) (5)	Secure data transmission systems referring to SMEs' requirements (11), (12), (13)
Technological and commercial exchange with leading countries in Industry 4.0 adoption (2)	Solutions technologically capable, but "downsized" to the needs of SMEs (12), (14)
Creation and implementation of systems to demonstrate associated technologies 4.0, applied to prioritized sectors (2) (4)	

(Continued)

TABLE 11.2 (continued)
Drivers for Digitalization in SMES

Brazil	Worldwide
Customization of existing solutions for different customers, from the most varied sectors (2) (4)	
Regulatory:	
Establishment of appropriate regulatory frameworks (1) (2) (5)	Industry-wide standards and approaches (12)
Establishment and promotion of open technical standards (interoperability) (1) (5)	

Source: (1) CNI 2016b, (2) CNI 2016a, (3) CNI 2019, (4) Stefanuto 2019, (5) BSA Foundation. Software.org. 2018, (6) ME 2020, (7) Horváth and Szabó 2019, (8) Chiarvesio and Romanello 2018, (9) Kiel et al. 2017, (10) Stentoft et al. 2020, (11) Hamzeh, Zhong, and Xu 2018, (12) Ghobakhloo and Fathi 2020, (13) Raj et al. 2020, (14) Müller, Buliga, and Voigt 2018

(Lehmann 2019), through knowledge transfer and training, accelerating the implementation process.

Secondly, it is vital the establishment of specific credit lines. "Hopes for inclusive and sustainable economic growth means helping to bring SMEs to the fourth industrial revolution," said Jeff Merritt, head of IoT, robotics, and smart cities at the World Economic Forum. Governments have the opportunity not only to reduce barriers to the adoption of new technologies but also to help position these companies for success in a rapidly evolving global economy (ME 2020).

11.7 INDUSTRY 4.0 OPPORTUNITIES

Industry 4.0 may be a game changer for SMEs. SENAI presents an example of how the adoption of new technologies in manufacturing brings competitive advantages to smaller companies. It is a pilot project with 43 companies from 24 Brazilian states. The project was the first to test in all regions of Brazil, the impact of the use of low-cost tools, such as sensing, cloud computing and IoT. Between May 2018 and October 2019, the project covered companies in the Food and Beverage, Metalworking, Furniture, Clothing and Footwear segments. SENAI specialists installed sensors, which collect data and store them. Then, the information is transmitted to the "*Minha Indústria Avançada*" platform (MInA; "My Advanced Industry"), which allows access to the data of the sensed machine. Through tablets and cell phones, managers can monitor, in real time, the performance of the production line and, with this, have greater control over process indicators and anticipate potential problems (CNI 2019). The digital technologies of Industry 4.0 increased the productivity of micro, small and medium-sized companies by an average of 22 percent. Although in a small sample, the project opens promising perspectives for SMEs. Table 11.3 presents the detailed results by region.

TABLE 11.3
Digitalization Impact on Productivity in 43 Brazilian SMEs

Region	improvement rate
Northeast	28.20%
Midwest	22.44%
North	22.29%
Southeast	18.42%
South	6.37%
Average	22.00%

Source: CNI 2019

The presented initiative (MInA) is focused on vertical integration, which leads us to intuit the potential benefits may be higher than the preliminary results (Table 11.3). As stated in (Stefanuto 2019), there are also new opportunities for digitalized SMEs, such as the Industry 4.0 as a priority on the Brazilian government's political agenda, the possibility of supply to multinationals, and internationalization.

Through several examples on the impact of Industry 4.0 on the triple bottom line, the barriers taken from the literature, although addressing mostly to the implementation of Industry 4.0, are equally relevant for leading to sustainable production. For example, when the access to credit is facilitated for SMEs, it increases the chances that the organizations will devote the funds to the activities supporting sustainable production practices. Similarly, addressing the barrier "disruption to existing jobs" looks after the social justice aspect of sustainable production (Khanzode et al. 2021).

Finally, sustainable implementation of digital technologies involves complex decision processes that permeate the design, planning and management and should be analyzed at the hierarchical levels: strategic, tactical, and operational. This enables stakeholders to make the right decisions in the Industry 4.0 implementation and plan to achieve the goals concerning sustainability in the long, medium, and short term (Aguiar et al. 2019).

11.8 CONCLUSION

SMEs are, on average, significantly less efficient than large companies. The new manufacture paradigm of Industry 4.0 is still incipient for most Brazilian companies. We presented the barriers and the drivers to a safe pathway towards Industry 4.0 in Brazil. However, digitalization also opens new perspectives to SMEs. Preliminary results point to an average increase of 22 percent in productivity. Also, the implementation of Industry 4.0 is a vital pathway to sustainable production.

REFERENCES

ABGI Brasil. 2018. *Guia de Legislação: Legislac¸a~o Relacionada à Inovação Tecnológica*. São Paulo. http://otd.cpqd.com.br/otd/wp-content/uploads/2018/11/ABGI-Legislac%CC%A7a%CC%83o-para-o-setor-de-tecnologia.pdf

Aguiar, Guilherme Teixeira, Gilson Adamczuk Oliveira, Kim Hua Tan, Nikolai Kazantsev, and Dalmarino Setti. 2019. "Sustainable Implementation Success Factors of AGVs in the Brazilian Industry Supply Chain Management." *Procedia Manufacturing* 39: 1577–1586. https://doi.org/10.1016/j.promfg.2020.01.284.

Bickman, Leonard, and Debra J. Rog. 2009. "The SAGE Handbook of Applied Social Research." *International Journal of Comparative Sociology* 19. https://doi.org/10.1163/156854278X00194.

Bowen, Glenn A. 2015. "Document Analysis as a Qualitative Research Method." *Qualitative Research Journal* 9 (2): 27–40. https://doi.org/10.3316/QRJ0902027.

BRASIL. 2018a. *DECRETO Nº 9.319 de 21 de Março de 2018*. Brasília, DF: Diário Oficial da República Federativa do Brasil. http://www.planalto.gov.br/ccivil_03/_ato2015-2018/2018/decreto/D9319.htm

BRASIL. 2018b. *Brazilian Digital Transformation Strategy*. Brasília, DF: E-Digital.

BRASIL. 2018c. *Documento de Referência Do Plano Nacional de Internet Das Coisas – Iot. Br*. Brasília, DF: MCTIC. http://otd.cpqd.com.br/otd/wp-content/uploads/2018/12/Cartilha-PLANO-NACIONALDE-INTERNET-DAS-COISAS_192x245_WEB.pdf

BRASIL. 2019. *DECRETO Nº 9.804 de 23 de Maio de 2019*. Brasília, DF: Diário Oficial da República Federativa do Brasil. http://www.planalto.gov.br/ccivil_03/_Ato2019-2022/2019/Decreto/D9804.htm

BRASIL, Ministério da Ciencia Tecnologia Inovações e Comunicações. 2017. *Plano de CT&I Para Manufatura Avançada No Brasil – ProFuturo*. Brasília, DF: MCTIC. https://antigo.mctic.gov.br/mctic/export/sites/institucional/tecnologia/tecnologias_convergentes/arquivos/Cartilha-Plano-de-CTI_WEB.pdf

BRASIL, Ministério da Ciencia Tecnologia Inovações e Comunicações, Centro de Pesquisa e Desenvolvimento em Telecomunicações CPqD, and Movimento Brasil Competitivo MBC. 2020. "Observatório Da Tranformação Digital." 2020.

BRASIL, Ministério da Ciencia Tecnologia Inovações e Comunicações, and Serviço Nacional de Aprendizagem Nacional SENAI. 2020. "Mapeamento 4.0." 2020.

BSA Foundation.Software.org. 2018. "Brazil 4.0: The Data-Driven Future of Brazilian Industries." https://software.org/wp-content/uploads/Software_Brazil4.0_English.pdf.

Chiarvesio, Maria, and Rubina Romanello. 2018. "Industry 4.0 Technologies and Internationalization: Insights from Italian Companies." In *International Business in the Information and Digital Age (Progress in International Business Research, Vol. 13)*, edited by R. van Tulder, A. Verbeke, and L. Piscitello, 357–78. Emerald Publishing Limited. https://doi.org/https://doi.org/10.1108/S1745-886220180000013015.

CNI. 2016a. "Desafios Para Indústria 4.0 No Brasil." *Confederação Nacional Da Indústria*, no. INDUSTRIA 4.0: 34. https://doi.org/2016.

CNI. 2016b. "Industry 4.0: A New Challenge for Brazilian Industry." *CNI Indicators* (April): 13. https://static-cms-si.s3.amazonaws.com/media/filer_public/13/e7/13e7e7bd-9b1d-4c16-8099–99b6d844d04e/special_survey_industry_40.pdf.

CNI. 2017. "Oportunidades Para a Indústria 4.0: Aspectos Da Demanda e Oferta No Brasil." *Confederação Nacional Da Indústria.* https://www.portaldaindustria.com.br/publicacoes/2018/2/oportunidades-para-industria-40-aspectos-da-demanda-e-oferta-no-brasil/

CNI. 2019. "Tecnologias Da Indústria 4.0 Aumentam Em 22%, Em Média, a Produtividade de PMEs." *Agência CNI de Notícias -Inovação e Tecnologia*. https://noticias.portaldaindustria.com.br/noticias/inovacao-e-tecnologia/tecnologias-da-industria-40-aumentam-em-22-produtividade-em-pequenas-e-medias-empresas/.

CNI. 2020. "As Invenções Da 4ª Revolução Industrial: Uma Análise Dos Dados de Patentes No Brasil." *Indústria 4.0* 1 (1): 1–14.

EdinburghGroup. 2014. "Growing the Global Economy through SMEs." *The Edinburgh Group*, 1–40. www.edinburgh-group.org/media/2776/edinburgh_group_research_-_growing_the_global_economy_through_smes.pdf.

Ghobakhloo, Morteza, and Masood Fathi. 2020. "Corporate Survival in Industry 4.0 Era: The Enabling Role of Lean-Digitized Manufacturing." *Journal of Manufacturing Technology Management* 31 (1): 1–30. https://doi.org/10.1108/JMTM-11-2018-0417.

Hamzeh, Reza, Ray Zhong, and Xun William Xu. 2018. "A Survey Study on Industry 4.0 for New Zealand Manufacturing." *Procedia Manufacturing* 26: 49–57. https://doi.org/10.1016/j.promfg.2018.07.007.

Horváth, Dóra, and Roland Zs Szabó. 2019. "Driving Forces and Barriers of Industry 4.0: Do Multinational and Small and Medium-Sized Companies Have Equal Opportunities?" *Technological Forecasting and Social Change* 146 (October 2018): 119–132. https://doi.org/10.1016/j.techfore.2019.05.021.

Kagermann, Henning, Wolfgang Wahlster, and Johannes Helbig. 2013. "Recommendations for Implementing the Strategic Initiative INDUSTRIE 4.0: Securing the Future of German Manufacturing Industry; Final Report of the Industrie 4.0 Working Group."

Kamble, Sachin S., Angappa Gunasekaran, and Rohit Sharma. 2018. "Analysis of the Driving and Dependence Power of Barriers to Adopt Industry 4.0 in Indian Manufacturing Industry." *Computers in Industry* 101 (June): 107–119. https://doi.org/10.1016/j.compind.2018.06.004.

Khanzode, Akshay G., P. R. S. Sarma, Sachin Kumar Mangla, and Hongjun Yuan. 2021. "Modeling the Industry 4.0 Adoption for Sustainable Production in Micro, Small & Medium Enterprises." *Journal of Cleaner Production* 279: 123489. https://doi.org/10.1016/j.jclepro.2020.123489.

Kiel, Daniel, Julian M. Müller, Christian Arnold, and Kai Ingo Voigt. 2017. "Sustainable Industrial Value Creation: Benefits and Challenges of Industry 4.0." *International Journal of Innovation Management* 21. https://doi.org/10.1142/S1363919617400151.

Lehmann, Ana Teresa. 2019. "Tendências Tecnológicas: Inserção Das PMEs Europeias Na Transformação Digital." *Secretária Especial da Produtividade, Emprego e Competitividade do Ministério da Economia do Brasil.* www.gov.br/economia/pt-br/centrais-de-conteudo/publicacoes/notas-informativas/2020/pmee0108_europeias-5-6-20.pdf.

ME. 2020. "Brasil Terá Primeiro Centro Afiliado Ao Fórum Econômico Mundial Focado Na Indústria 4.0." Ministério Da Economia. 2020. www.gov.br/economia/pt-br/assuntos/noticias/2019/11/brasil-tera-primeiro-centro-afiliado-ao-forum-economico-mundial-focado-na-industria-4.0.

Müller, Julian Marius, Oana Buliga, and Kai Ingo Voigt. 2018. "Fortune Favors the Prepared: How SMEs Approach Business Model Innovations in Industry 4.0." *Technological Forecasting and Social Change* 132 (September 2017): 2–17. https://doi.org/10.1016/j.techfore.2017.12.019.

Raj, Alok, Gourav Dwivedi, Ankit Sharma, Ana Beatriz Lopes de Sousa Jabbour, and Sonu Rajak. 2020. "Barriers to the Adoption of Industry 4.0 Technologies in the Manufacturing Sector: An Inter-Country Comparative Perspective." *International Journal of Production Economics* 224 (August 2019): 107546. https://doi.org/10.1016/j.ijpe.2019.107546.

Schaefer, J. L., I. C. Baierle, M. A. Sellitto, J. C. M. Siluk, J. C. Furtado, and E. O. B. Nara. 2020. "Competitiveness Scale as a Basis for Brazilian Small and Medium-Sized Enterprises." *Engineering Management Journal*, 1–17. https://doi.org/10.1080/10429247.2020.1800385.

SEBRAE. 2014. "Participação Das Micro e Pequenas Empresas Na Economia Brasileira." *SEBRAE Biblioteca*, 106.

Stefanuto, Giancarlo Nuti. 2019. "Tendências Tecnológicas: Inserçao Das PMEs Brasileiras Na Transformaçao Digital." *Secretária Especial da Produtividade, Emprego e Competitividade do Ministério da Economia do Brasil.* www.gov.br/economia/pt-br/centrais-de-conteudo/publicacoes/notas-informativas/2020/pmee0108_brasileiras-5-6-20.pdf/view.

Stentoft, Jan, Kent Adsbøll Wickstrøm, Kristian Philipsen, and Anders Haug. 2020. "Drivers and Barriers for Industry 4.0 Readiness and Practice: Empirical Evidence from Small and Medium-Sized Manufacturers." *Production Planning and Control*, 1–18. https://doi.org/10.1080/09537287.2020.1768318.

UNIDO. 2020. "Competitive Industrial Performance Report." Vol. II. https://stat.unido.org/country-profile.

The World Bank. 2020. "Manufacturing, Value Added (% of GDP)." https://data.worldbank.org/indicator/NV.IND.MANF.ZS.

WTO. 2016. "World Trade Report 2016." *World Trade Organization* 7. https://doi.org/10.1017/S1474745608004035.

12 Cloud-Based SMART Manufacturing in China

Y. C. Chau and Vincent W. C. Fung

CONTENTS

12.1 INTRODUCTION

Industry 4.0 is a term often used to refer to the developmental process in the management of manufacturing and chain production. The term also refers to the fourth Industrial Revolution. Industry 4.0 was first publicly introduced in 2011 as "Industrie 4.0" by a group of representatives from different fields (such as business, politics, and academia) under an initiative to enhance the German competitiveness in the manufacturing industry. The German federal government adopted the idea in its High-Tech Strategy for 2020. Subsequently, a working group was formed to further advise on the implementation of Industry 4.0. In October 2012, the working group on Industry 4.0 presented a set of Industry 4.0 implementation recommendations to the German federal government. The workgroup members and partners are recognized as the founding fathers and driving force behind Industry 4.0. On 8 April 2013 at the Hannover Fair, the final report of the working group on Industry 4.0 was presented (Forschungsunion acatech 2013).

COUNTRY BACKGROUND

China started the open-door policy in 1978, and under the fast growth of the manufacturing industry, it has faced a growth limitation since turn of the millennium.

DOI: 10.1201/9781003165880-12

According to State Council of the People Republic of China, Made in China 2025, July 7 2015, Prime Minister Li Keqiang launched "Made in China," an initiative that seeks to modernize China's industrial capability, which is mainly focused on the digital manufacturing transformation.

12.2 DIGITALIZATION POLICY IN THE COUNTRY

12.2.1 Made-in-China 2025

China revealed its plan to become a world manufacturing power by 2025 on 19 May 2015. According to Xinhua News Agency, the "Made in China 2025" plan, endorsed by Premier Li Keqiang, "Is the country's first action plan focusing on promoting manufacturing."

However, there are big gaps among Chinese enterprises; some of them still need to develop from industry 2.0 to industry 3.0. Miao Wei (China's industry and information technology minister) said the Made in China 2025 strategy is an overall plan for transforming and upgrading the manufacturing industry.

It may sound odd that China, often called the "world's factory," is focusing such high-level attention on strengthening its manufacturing capabilities. But China's leaders want to move up the supply chain – they want China to become known as a leader in manufacturing innovative technologies. In Li's words, Beijing wants to go from "Made in China" to "Created in China."

The action plan also calls for restructuring of the manufacturing sector, to remove overcapacity that exists in key industries like steel. Here, China faces a dilemma: the only surefire way to solve the problem is to shut factories down, which would result in a spike in unemployment (something Beijing tries its hardest to avoid).

While China is shifting away from certain industries, Beijing wants to move towards others. The ten-year plan calls for "promoting breakthroughs in 10 key sectors." Those key sectors (and their relationship to broader strategic goals) are as follows:

1. Information technology: Beijing has emphasized the need to wean China off a reliance on foreign technology, and for China to become a "cyber power" in its own right.
2. Numerical control tools and robotics: Becoming a leader in numerical control machines and robotics will help China keep its manufacturing industry strong even as labor costs rise relative to other developing countries.
3. Aerospace equipment: This is a prestige symbol for China, which also has security implications.
4. Ocean engineering equipment and high-tech ships: China is also hoping to become a technological leader in this field.
5. Railway equipment: China has made a name for itself by building an astonishing number of railways, both at home and abroad. Beijing clearly wants to keep its competitiveness in this area as it continues to push infrastructure construction through the "Belt and Road" initiative.

6. Energy saving and new energy vehicles: China's "war on pollution" meshes well with a long-standing ambition to win global recognition for Chinese-made cars. Beijing is betting that energy-efficient and clean energy vehicles will be the wave of the future.
7. New materials: The holy grail of advanced manufacturing technologies as well as strong research and development capabilities. If China can become a leader in turning new materials into useful products, it will have arrived on the innovation stage in a big way.
8. Medicine and medical devices: Another area where China seeks international prestige, plus less reliance on foreign companies for necessary equipment and supplies.
9. Agricultural machinery: In addition to the potential for exports, agricultural machinery is of crucial importance to China itself as it seeks to maximize efficiency in its agricultural sector.
10. Power equipment: Although not made explicit, here again energy efficiency is the key goal. China is actively pursuing smart grid and smart city technologies.

Clearly, China's manufacturing sector needs a push to take it to the next level, and this is where Made-in-China 2025[1] comes in. According to the government's action plan, China will aim at a big leap in innovation as well as manufacturing efficiency and realize basic industrialization by 2025; being able to compete with developed manufacturing powers by 2035; and leading the world's manufacturing by the 100th birthday (2049) of the New China. The plan will focus more on high-tech industries such as information technology, robotics, aerospace, and new materials. But clearly, there's a long way to go.[2]

12.3 MOTIVATION FOR ADOPTING INDUSTRY 4.0 IN SMEs

In many SMEs, two major challenges evince the most problems in the daily manufacturing operation. They are strategic cost performance (SCP) and quality yield performance (QYP). These two challenges become huge barriers to SMEs moving forward in business expansion plans. By contrast, such challenges magnify into a significant driving force to motivate SMEs to improve productivity and cost efficiency, and in return achieve a lower cost benefit and a higher quality performance optimization. Both uplift the SME into the next level of success and eventually meld into one of the core competencies for the company.

12.3.1 Strategic Cost Performance (SCP)

Production systems run as a stand-alone entities or silos without interconnectivity and interpretability. Data collection becomes difficult in the production chain, which

1 http://english.www.gov.cn/2016special/madeinchina2025/ (accessed 29.05.2021)
2 www.hkengineer.org.hk/issue/vol44-may2016/cover_story/ (accessed 29.05.2021)

causes many negative snowball effects along the subsequent work processes. If a defect is found at a later stage of the production cycle, it is costly and time consuming to reverting it back to an acceptable criterion for quality conformity. Production lots may have to be rebuilt and remanufactured. Time to market is unpredictable and result in late shipment to customers. Material scrap and time waste create loss of sale and negative profit margins to SMEs.

Factories must identify the faults immediately so as not to keep the production line idle for long. Individual production systems may be needed to correct back to an ultimate parameter setting prior to giving a green light for production to resume. Production processes are normally running at a dynamic mode and continuously performing at full speed of capacity from pieces to pieces and batches to batches. Production quantities built up as "works-in-process" or finished items are of concern if defects are found unexpectedly during such production periods.

Strategic cost performance (SCP) comprises two primary constituents that influence most the financial performance of a SME. They are "executional cost management" and "structural cost management." According to (Henri, Boiral, & Roy, 2016), executional cost management is a short-term tactic. It is an operational strategy to deal with environmental aspects through cost reduction process such as redundancy and overlapping hierarchy layers.

Structural cost management is a long-term strategy in relation to product design, production stability, and process yield efficiency. Product design is directly related to complexity of the product structure, material selection, material utilization, and product compliance by the designated country. Production stability involves machine health maintenance and machine deterioration, material continuity due to supply chains, and material deviation. Process yield efficiency relates to production optimization, process variation that leads to quality deficiency, human error, equipment failure, and frequency of false calls along the production system.

12.3.2 Quality Yield Performance (QYP)

Inconformity of quality specification, customer returns, and remanufacturing of production turn a SME into a tremendous financial impact. The time to recall the problem from a customer complaint sometimes takes too long to be remedied. In many circumstances, the SME loses the "golden time" to resolve the issues effectively and efficiency at a minimum time, effort, and cost. In consequence, production assembly is halted, and workers and production machines are idle. Production yield or process yield is badly affected due to quality uncertainty.

The economic impact from loss of productivity in an SME is difficult to measure tangibly and precisely. Cost of production interruption is defined as single disruption and concurrent disruption in a multistage manufacturing system (Bai, Kajiwara, & Liu, 2016). Single disruption is normally calculated by the active-based costing (ABC) method, while concurrent disruptions often appear simultaneously at various workstations as well as multiple production systems. The cost for a single-stage production disruption is liner while a multistage manufacturing system is exponential from one affecting other workstations towards the end of the production chain. The accumulated disruption cost for both single- and multistage production become a tremendous financial loss to the company.

To summarize the motivation as a driver for Industry 4.0, cost and quality performance are the leading criteria of sustainability and profitability in the manufacturing industry for a SME. SCP and QYP are vital, in particular, at a multistage production chain such as SMT manufacturing systems in the electronics manufacturing services (EMS) industry. Industry 4.0 architecture, or a "smart manufacturing" paradigm shift is crucial to drive production costs into a more competitive advantage and to push quality performance levels onto a new horizon.

12.4 SMT IN ELECTRONICS MANUFACTURING SERVICES (EMS)

According to the research, the total value of the electronics assembly worldwide was estimated to US$1.3 trillion in 2018. This is expected to grow to approximately US$1.5 trillion in 2023 and shows that the market potential of electronics assembly is relatively high, in which EMS are regarded as the determinant force in electronics production, accounting for 42 percent of all the assembly tasks. EMS are also deemed to be the most desired manufacturing model in the world for original equipment manufacturers (OEMs), and therefore, the market is expected to grow from US$542 billion in 2018 to US$777 billion in 2023.

Due to the substantial growth in production of the aforementioned electronic products, the requirements for product quality and productivity of EMS have become stricter and more complicated in recent years (Barrad & Valverde, 2020). On one hand, EMS companies have to design, manufacture, test, distribute, and provide field return services, in which small and precise electronic components are populated onto the printed circuit board (PCB). On the other hand, customers have become more demanding in relation to product quality and reliability throughout the entire product life cycle (PLC). Industry 4.0, or the smart manufacturing paradigm, is the best tool to connect the entire PLC from product design, manufacture, testing, distribution, after sales, product quality, and reliability of life history together for product integrity and for future product design generation.

In view of these considerations, EMS companies are looking for improvements in manufacturing capabilities, cost effectiveness, productivity performance, and product quality assurance. In addition to the current electronics manufacturing industry, as manufacturing technology continues to flourish, production maneuvering in the factory is being shaped into Industry 4.0. Production mixes in multiple dimensions of "High volume-low variety" and "Low volume-high variety" become essential in maintaining business competitiveness. Smart manufacturing, or Industry 4.0, facilitates the advantage of economic production runs and strategic supply chain management for mass customization and personalization of manufacturing paradigms (Mourtzis, 2016).

For the substantial market demand of electronic items, production of PCBA (printed circuit board assembly) is deemed to be a critical process in manufacturing the designated PCBs by means of the appropriate assembly method, such as through-hole technology (THT) and SMT (Lau et al., 2016). Currently, most of the THT operations have been replaced by SMT, which is suitable for producing high volumes of products of a greater level of quality requirement. With the adoption of Industry 4.0 in SMT production processes, the yield performance and product quality of the finished PCBA are more reliable and stable for the EMS companies. As shown in Figure 12.1, the generic PCBA process using SMT is illustrated, and there are ten major multi-stages in the workflow, namely (1) PCB cleaning, (2) solder paste printing, (3) solder paste inspection (SPI), (4) modular SMT, (5) reflow oven, (6) buffering, (7) automatic optical inspection (AOI), (8) in-circuit tester (ICT), (9) PCB depanelizer, and (10) storage and distribution of finished PCBs. With regard to the SPI, AOI, and ICT, these are added in the PCBA process to inspect the quality of PCBs during the surface mount process (Dong et al., 2012). Therefore, the quality of the finished PCBs can be guaranteed, while the entire production flow can operate in an automatic manner from the bare PCBs to the mounted PCBs.

Apart from enhancing operations management, the PCB design does not consider the process performance of PCBA and capabilities of manufacturing sites, since the knowledge of critical factors in PCBA and machine parameter settings cannot be shared effectively. Subsequently, the PCB designs from the Gerber drawing may not be suitable for the specific manufacturing sites, in which process performance is relatively uncertain in relation to the EMS companies and clients. Therefore, an effective measure to estimate process performance and manufacturing capabilities

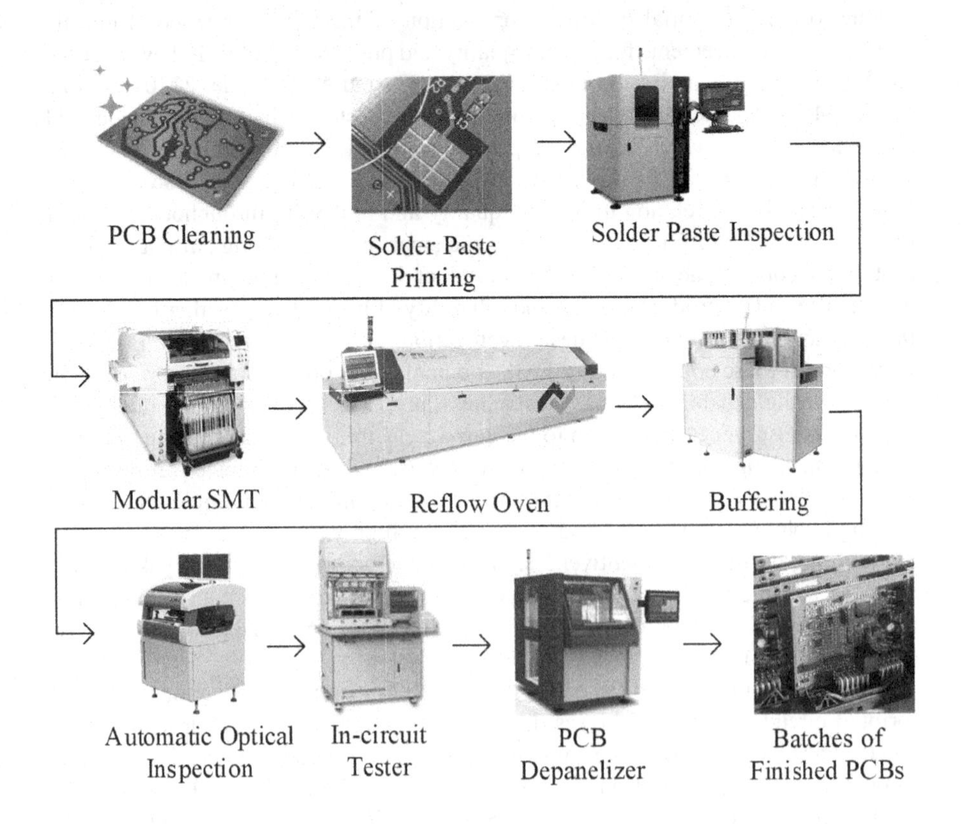

FIGURE 12.1 Generic SMT-based PCBA process.

should be developed at the planning stage of designing PCBs. In addition, effective measures of knowledge transfer in the manufacturing industry are lacking (Chuang, 2013). Valuable work experience can only be passed on gradually from one person to another. Manufacturing knowhow is strictly self-reliant, unstructured, and limited to certain close team members. Thus, manufacturing knowledge is difficult to transfer which should be the norm in the manufacturing electronic industry.

In view of these considerations, an intelligent manufacturing performance predictive framework is suggested to improve the EMS's capabilities, cost efficiency and quality performance. To demonstrate the benefit of implementing Industry 4.0 into SMT work process, a comprehensive research study was carried out in the case factory.

12.5 INDUSTRY 4.0 ADOPTED IN SMT SHOP FLOOR

Adopting the techniques in design of experiment (DoE) and artificial intelligence (AI), the intelligent manufacturing performance predictive framework (IMPPF) is designed in a modular format as illustrated in Figure 12.2. It consists of three major

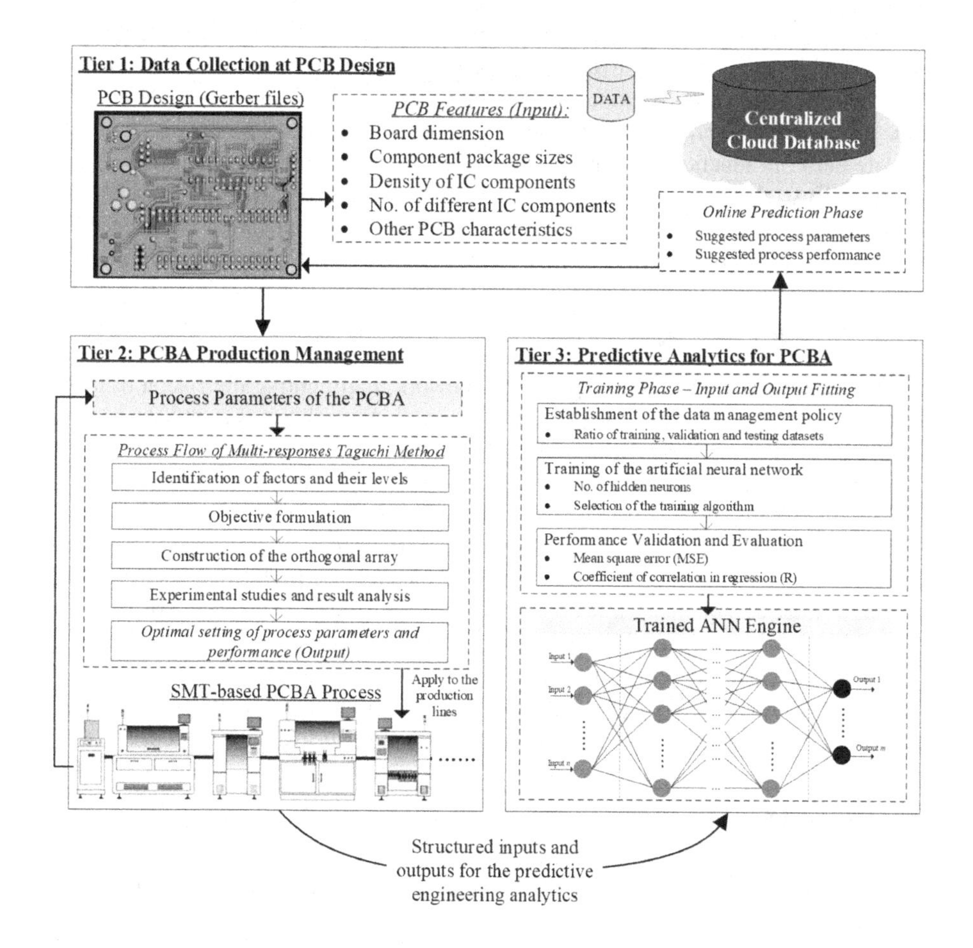

FIGURE 12.2 Modular framework of the IMPPF.

tiers: (1) data collection at PCB design, (2) PCBA production management, and (3) predictive analytics for PCBA. A closed-loop management for the PCB ecosystem from the design stage to the production stage is established, where the PCB features, process parameters, and process performance are integrated together, so as to build predictive engineering analytics in the smart manufacturing environment.

12.6 INDUSTRY 4.0 OPPORTUNITIES

With the adoption of Industry 4.0, intelligence can be given to the PCBA process to fine tune process parameter settings and improve process performance, while predictive analytics can be obtained to estimate the quality performance of the PCBA process, starting with the Gerber drawing. Industry 4.0 establishes a closed-loop management for the entire lifecycle of the PCBA, in which operational data can be utilized to formulate the predictive engineering analytics. With the collection of the operational records using the IMPPF in Industry 4.0, a reliable data source can be managed, which can be further used to formulate predictive engineering in the PCBA process.

This can facilitate design and production planning before the start of PCB manufacturing to estimate optimized process parameters and process performance. Consequently, the design, production planning, and PCBA can be considered as a whole to maximize process performance. In addition, the EMS companies can formulate better production planning, cost control and quality management in relation to the PCBA, eliminating waste from the production process and optimizing yield performance. Therefore, the knowledge transfer and communication between PCB design and production can be integrated to avoid the mismatch of information.

12.7 CONCLUSION

IMPPF in Industry 4.0 is an integrated framework, highlighting the fusion of techniques in DoE and AI, while the predictive engineering analytics is established in the PCBA process to facilitate the closed-loop management. In the closed-loop management, PCB design and production can be integrated to generate a synergy in predicting and improving process parameters and quality performance for the PCBA in the EMS industry. Therefore, the PCBA in the EMS industry can be further developed in relation to Industry 4.0. To restore China's outstanding performance position in cost advantage and quality while competing with other Asian countries, SMEs in the manufacturing industry need to adopt Industry 4.0. To further leverage the advantage of Industry 4.0 via implementing Industrial Internet Of Things initiatives, Chinese SMEs can connect the supply chains and integrate production operations seamlessly in various manufacturing locations as a jointed manufacturing ecosystem.

REFERENCES

Bai, G., Kajiwara, T., & Liu, J. (2016). Measuring the Cost of Individual Disruptions in Multistage Manufacturing Systems. *Journal of Management Accounting Research*, 28(1), 1–26. doi:10.2308/jmar-50924.

Barrad, S., & Valverde, R. (2020). The Impact of e-Supply Chain Management Systems on Procurement Operations and Cost Reduction in the Electronics Manufacturing Services Industry. *Journal of Media Management and Entrepreneurship*, 2(1), 1–27.

Chuang, Y. S. (2013). Learning and International Knowledge Transfer in Latecomer Firms: The Case of Taiwan's Flat Panel Display Industry. *IEEE Transactions on Engineering Management*, 61(2), 261–274.

Dong, N., Wu, C. H., Ip, W. H., Chen, Z. Q., & Yung, K. L. (2012). Chaotic Species-Based Particle Swarm Optimization Algorithms and Its Application in PCB Components Detection. *Expert Systems with Applications*, *39*(16), 12501–12511.

Forschungsunion acatech (2013). Recommendations for implementing the strategic initiative INDUSTRIE 4.0, Final report of the Industrie 4.0 Working Group, 08 April 2013

Henri, J.-F., Boiral, O., & Roy, M.-J. (2016). Strategic Cost Management and Performance: The Case of Environmental Costs. *The British Accounting Review*, 48(2), 269–282. doi:10.1016/j.bar.2015.01.001.

Lau, C. S., Khor, C. Y., Soares, D., Teixeira, J. C., & Abdullah, M. Z. (2016). Thermo-Mechanical Challenges of Reflowed Lead-Free Solder Joints in Surface Mount Components: A Review. *Soldering & Surface Mount Technology*, 28(2), 41–62.

Mourtzis, D. (2016). Challenges and Future Perspectives for the Life Cycle of Manufacturing Networks in the Mass Customisation Era. *Logistics Research*, 9(1), 2.

13 Industry 4.0 and SMEs
An Indian Perspective

Debabrata Das, Sankhadip Kumar, and Amit Somwanshi

CONTENTS

13.1 INTRODUCTION AND COUNTRY BACKGROUND

India remains as one of the rapidly growing major economies of the world driven by a surge in domestic consumption and investment growth in recent years. The importance of micro-, small, and medium-sized enterprises (MSMEs) in driving economy is mainly due to the fact that it forms 33.4 percent of India's manufacturing output and around 8 percent of the total GDP through production, export, and employment generation (GOIMSME, 2016). As per the revised classification applicable from 1 July 2020, a composite criteria has been formed stating that companies with capital expenditure not more than INR 10 million and annual revenue not more than INR 50 million are micro; companies with investment in capital expenditure not more than INR100 million and annual revenue not more than INR 500 million are small; and companies with investment in capital expenditure not more than INR 500 million and annual revenue not more than INR 2.5 billion are medium enterprises.[1] In the present section, we study how Indian manufacturing SMEs perceive Industry 4.0, and what are the drivers and

[1] Corresponding author Email ID: debabrataiitb@gmail.com (D. Das)
Before 01-July-2020, enterprises with investment < INR 2.5 million classified as micro, investment < INR 50 million as small and investment < 100 million as medium.

DOI: 10.1201/9781003165880-13

challenges of adopting Industry 4.0 in their enterprise as far as collaborative value creation among the supply chain partners is concerned.

13.2 DIGITALIZATION POLICY IN INDIA

In a bid to push wider application of Industry 4.0 across the SME sector of India, the government of India has not only encouraged foreign direct investment but also brought in several favorable policies. The Ministry of Electronics and Information Technology (MEIT, 2015) prepared a draft Internet of Things (IoT) policy to develop IoT products in all possible domains, and to undertake various capacity (both human and technology) development programs for improving IoT-specific skill sets. The Centre of Excellence for IoT, as a part of the "Digital India Initiative" was announced by the prime minister of India in July 2015 to kickstart the IoT ecosystem for reducing import dependency on IoT components and to promote indigenization (CEIoT, 2015). Smart Advanced Manufacturing and Rapid Transformation Hub (SAMARTH), Udyog Bharat 4.0, is an initiative of Ministry of Heavy Industry and Public Enterprises, Government of India, which aims at raising understanding about Industry 4.0 technologies among Indian manufacturing industries through its experiential and demonstration centers. To help the growing need of Industry 4.0, the government of India has also started expanding its flagship "Skill India" mission to include artificial intelligence, the IoT, and other emerging technologies by keeping in mind that 40 percent of India's total workforce must be up-skilled over the next five years in cutting-edge technologies (NASSCOM report, 2018). Not only the government but also the private information technology companies have stepped in for a global drive for technology re-skilling of more than one million professionals on the World Economic Forum's SkillSET portal by January 2021.

13.3 METHODOLOGY

13.3.1 Case Selection

Despite continuous efforts of public and private stakeholders over the years, there has been only a little boost in the actual scenario in terms of implementation of Industry 4.0 in Indian SMEs. To understand the reasons behind the low adoption rate of Industry 4.0 technologies, this study was carried out by interviewing executives from a few SMEs. After exploring the database of the All India Industries Association, 14 firms were carefully assessed, and three manufacturing firms were selected from diverse industries for this study as the companies satisfied the definitional framework of SMEs and had adopted some Industry 4.0 technologies in their operations.

Selected firms were located in one of the major industrial clusters of Maharashtra, a western state of India, and all had more than 25 years of experience in diverse industries ranging from automotive, agricultural, off-highway, printing, and packaging.

13.4 PROFILE OF SELECTED FIRMS

Case Study Firm 1: It was founded in the year 1982 by an Indian entrepreneur. The firm started its commercial production in an industrial cluster of western India. The enterprise specializes in manufacturing propeller shafts, double cardan shafts, universal joints, and components. They supply their products to various automotive, agricultural, off-highway, industrial, and aftermarket companies. They have been using industrial robots to reduce machine breakdown and delivery time and to improve quality of the final products.

Case Study Firm 2: The enterprise was founded in the year 1990 by two entrepreneurs in one of the major industrial cluster of Maharashtra, a state in western India. The core activities of the firm are manufacturing of special purpose machine for printing and packaging industry. They adopted Industry 4.0 technologies such as 3D printing, industrial robots, and IoT sensors to improve productivity and enhance quality of the final product.

Case Study Firm 3: The enterprise was founded in the year 1993 by three Indian entrepreneurs in one of the major industrial cluster of Maharashtra, a state in western India. They provide turnkey solutions to automotive industry. Their products include welding guns, microprocessor weld timers, SPMs, transformer for portable guns, spot welding electrodes, high conductivity copper alloy castings, and forgings. They adopted Industry 4.0 technologies such as 3D scanner, industrial robots, and additive manufacturing to reduce delivery time and improve productivity.

The demographic information of these three firms and the details regarding the Industry 4.0 implementation is summarized in Table 13.1.

TABLE 13.1
Demographic Information and Industry 4.0 Technologies of the Case Study Firms

Firm	Age (in years)	Industry & main activities	Number of Employees	Annual Turnover (in USD)	Industry 4.0 technologies adopted	Main purpose of Industry 4.0 technologies
Firm1	37	Machining (Propeller shaft manufacturing)	800	~ 30 million	– Industrial robots	– Produce quality products – Reduce delivery time – Reduce breakdown of machines
Firm2	29	Manufacturing of special purpose machine for printing and packaging industry	100	~ 4 million	– Additive manufacturing – Industrial robots – IoT sensors – Cloud ERP	– Improve productivity – Improve quality

(*Continued*)

TABLE 13.1 (continued)
Demographic Information and Industry 4.0 Technologies of the Case Study Firms

Firm	Age (in years)	Industry & main activities	Number of Employees	Annual Turnover (in USD)	Industry 4.0 technologies adopted	Main purpose of Industry 4.0 technologies
Firm3	26	Automotive Industry (sheet metal stamping for automobile parts)	500	~ 31 million	- 3D scanner - Industrial robots - Additive manufacturing	- Sustainable growth - Increase profitability - Improve productivity

13.5 DATA COLLECTION

In the current study, we performed qualitative research by collecting primary data from the respondents using a structured questionnaire, which was floated to senior-level executives from the enterprises. After receiving the online responses, the interviewer had telephonic conversations with the respondents to validate the data and bring rigor in the data collection process. Since the interviewer knows the interviewee personally, we believe that the validity of the results is not affected by using phone as a medium of interview (Blome and Schoenherr, 2011). The open-ended questions captured information on business demographics, business characteristics, and the details of Industry 4.0 technologies adopted. The questions are listed in Appendix A whereas the interview was focused to know about the implementation experience of Industry 4.0 technologies throughout the firm (downstream, vertical, and upstream) to understand the challenges and barriers.

13.6 RESULTS AND DISCUSSION

After carefully analyzing the data and discussing with senior executives of the SMEs, it was found that productivity and efficiency stand out to be the primary objectives of implementing Industry 4.0 in these three enterprises. Therefore, it was of no surprise that the majority of their investment on Industry 4.0 technologies had been directed towards implementing industrial robots and additive manufacturing across the production line which was in line with their objectives. Key point to note that firm 2 also invested on technologies like cloud computing, IoT sensors to increase information sharing, and information technology security in a bid to move to a more efficient process and gain an upper hand while exploring new market in the domain of printing and packaging industry.

Regarding challenges, it was observed that most of the firms cited the lack of standardization or knowledge about the data format of the suppliers as the major and common challenge for horizontal upstream integration (with the suppliers) of Industry 4.0 technologies. In some cases, it was argued that the insights gained from suppliers' real-time data seemed to be over-engineering for the business of SMEs. Meanwhile, from the customers' perspective, it was observed that data-driven business model is still not the priority for the industrial and end customers of SMEs, which deters the company from investing in technologies for establishing digital interconnection in the downstream of the value chain. Too much information about the product and price can also increase the bargaining power of customers and original equipment manufacturers (OEMs), which may result in losses for the SMEs implementing Industry 4.0 technologies in the downstream.

For vertical integration within the organization, the key barrier for implementation of Industry 4.0 technologies is the cost of implementation, as many SMEs tend to be risk averse. In some cases, hand labor is preferred over industrial robots as the former is more flexible and cheaper for mass production. Most of the firms surveyed stated the inability to attract skilled and qualified experts to be one of the major hindrances in implementing cutting-edge technologies across various interfunctional departments.

Finally, to leverage the full power of Industry 4.0 and to overcome the aforementioned challenges, an organization needs to inculcate digital strategies where the primary focus should be on change management and digitization which is lacking at the moment. With this, a culture of innovation can also be built that can drive the companies towards process improvement.

13.7 INDUSTRY 4.0 DRIVERS & OPPORTUNITIES IN INDIA

Empirical studies on hundreds of SMEs from several countries have revealed key factors that drive application of Industry 4.0 technologies in SMEs related to technology domain, company domain, and industry domain. Degree of similarity, knowledge about technology, and effect on flexibility feature among factors related to technology domain. Company-related factors consist of drivers like company size, strategy rationalization process, and Industry 4.0 readiness. Lastly, drivers like characterization, suitability, and culture for innovation of SMEs in implementation of advanced manufacturing technologies, regulatory policy peer pressure from industry, and globalization strategies have been considered under industry-related factors.

In the current business environment, SMEs in India are looking to use Industry 4.0 technologies to satisfy the global quality standards and efficiency, which can position them better to compete with multinational enterprises in a sustainable way. To improve quality, additive manufacturing is shown to have contributed by designing products with greater accuracy. Industry 4.0 technologies like industrial robots in manufacturing industry facilitates machine to machine (M2M) interaction, which leads to better plant optimization and productivity. Integration of cloud into ERP systems provides a platform for real-time data collection with cost-effective data storage for Indian SMEs.

13.8 INDUSTRY 4.0 BARRIERS IN INDIA

Regarding challenges, it was observed that 83 percent of the respondents cited the lack of standardization or knowledge about the data format of the suppliers as the major and common challenge for horizontal upstream integration (with the suppliers) of Industry 4.0 technologies. In some cases, it was argued that the insights gained from suppliers' real-time data seemed to be over-engineering for the business of SMEs. Meanwhile, from the customers' perspective, it was observed that the data-driven business model is still not the priority for the industrial and end customers of SMEs, which deters the company from investing in technologies for establishing digital interconnection in the downstream of the value chain. Too much information about the product and price can also increase the bargaining power of customers and OEMs, which may result in financial losses for the SMEs implementing Industry 4.0 technologies the downstream.

13.9 INDUSTRY 4.0 IN SMEs: INDIA VS. GLOBAL

Indian SMEs in the manufacturing sector are investing more in technologies like industrial robots and additive manufacturing to improve productivity and efficiency unlike their counterparts in developed countries like Denmark and Germany where SMEs have more interest in Industry 4.0 technologies related to information systems and communication technology. This is possibly because companies in developed countries assume that their existing technologies are superior to external newer ones since they have been developed inhouse (Katz and Allen, 1982). One more challenge that is common among German SMEs is that four out of ten SMEs do not have a digital strategy for Industry 4.0 (Schröder, C., 2016). This is less of a challenge in the Indian landscape since the technology maturity level is still at a nascent stage. However, the lack of standards and poor data security regarding interface technologies remain a challenge for SMEs in both Germany and India. A detailed study of the SMEs from the North Sea region in Europe highlights opportunities in the areas of productivity, revenue growth, employment, and investment and also reveals lack of business support, overcautious nature, shortage of skilled workforces, and a lack of standards and security to be the major obstacles for adoption of Industry 4.0 technologies. A similar study conducted among SMEs in UK (Masood, T. and Sonntag, P., 2020) suggests that technologies like additive manufacturing, artificial intelligence, and virtual reality stand out among others due to their high benefits and low complexity where technology like additive manufacturing is found to be common from Indian perspectives as well. It has also been seen during this study that company size affects the benefits gained by an SME implementing Industry 4.0 technologies, which is true in the Indian context as well, where adoption of these technologies has been majorly limited to medium and larger firms where financial resources are available, knowledge and skills are present, and technology awareness is prevalent among employees. From the foregoing discussion, it can be seen that opportunities and challenges identified in different studies are found to be more or less comparable to that of the Indian scenario, and it can be concluded that these are generic and global in nature.

13.10 DISCUSSION & CONCLUSION

As per the available data from the Central Statistics Office (CSO), Ministry of Statistics & Programme Implementation, the share of MSME in GDP stands at 28.9 percent, which has dipped marginally over the years despite favorable government policies in place. In such a scenario, the early adoption of digital technologies like Industry 4.0 can be the gamechanger for sustainable growth in this sector. As per the study, it was observed that such companies have improved their production efficiency and quality through investments in technologies like additive manufacturing and industrial robots and have reduced IT costs by leveraging cloud ERP models. It is also believed that available cheap labor and the change in culture majorly driven by supporting policies will further improve the rate of adoption among Indian SMEs. However, much of the aforementioned benefits have so far been met with initial challenges like a lack of knowledge about standardization and the high cost of product development. These challenges can be mitigated by a skilled workforce and by having a proper digital strategy in place. While comparing the findings to the global standards, it has also been seen that the trend regarding technologies, benefits, and challenges for MSMEs are aligned with that seen in other parts of the world.

REFERENCES

Blome, C., & Schoenherr, T. (2011). Supply chain risk management in financial crises – a multiple case-study approach. *International Journal of Production Economics*, *134*(1), 43–57.

Government of India, Centre of Excellence for IoT (CEIoT). (2015). Retrieved from www.ernet.in//projects/iot.html.

Government of India, Ministry of Micro, Small and Medium Enterprises (GOIMSME). (2016). Annual report 2015–16. Retrieved from https://msme.gov.in/sites/default/files/MEME%20ANNUAL%20REPORT%202015-16%20ENG.pdf.

Katz, R., & Allen, T. J. (1982). Investigating the Not Invented Here (NIH) syndrome: A look at the performance, tenure, and communication patterns of 50 R & D Project Groups. *R&d Management*, *12*(1), 7–20.

Masood, T., & Sonntag, P. (2020). Industry 4.0: Adoption challenges and benefits for SMEs. *Computers in Industry*, *121*.

Micro, Small & Medium Enterprises Development (MSMED) Act, 2006. Government of India. Government of India, Ministry of Electronics and Information Technology (MEIT). (2015). Retrieved from https://meity.gov.in/content/revised-draft-internet-thingsiot-policy-0.

NASSCOM Report. (2018). Retrieved from www.nasscom.in/knowledge-center/publications/nasscom-artificial-intelligence-primer-2018.

Schröder, C. (2016). *The Challenges of Industry 4.0 for Small and Medium-Sized Enterprises*. Bonn, Germany: Friedrich-Ebert-Stiftung.

Appendix A: Questions asked to SMEs

Q1. For how long has the enterprise been operational?
Q2. What are the core activities of your enterprise?
Q3. How many employees are working in your enterprise?
Q4. What is the annual turnover?
Q5. What are the Industry 4.0 technologies adopted by your enterprise?
Q6. For what purposes are these technologies being used?
Q7. What are the outcomes achieved by adopting these technologies?

Appendix B

TABLE B1
Overview of experience of Industry 4.0 implementation of Firm 1

Challenges / Barriers	Horizontal Integration (Upstream)	Vertical Integration	Horizontal Integration (Downstream)	End-to-End Engineering
High Investment in Industry 4.0 Implementation	Lack of a clear business case	Lack of a clear digital operations vision and support		
Lack of clarity regarding the economic benefit				High cost of data storage
Challenge in value chain integration		Difficulty in coordinating actions		
Risk of security breaches				
Low maturity level of the desired technology				No data consistency across the lifecycle of products
Inequality				
Disruption to existing jobs		Old, long-serving workforce fears to be replaced		

Challenges / Barriers	Horizontal Integration (Upstream)	Vertical Integration	Horizontal Integration (Downstream)	End-to-End Engineering
Lack of standards, regulations and forms of certification	Inappropriate data formats from suppliers			
Lack of infrastructure				
Lack of digital skills	No such digitization solutions			Less requirement of data storage facilities
Challenges in ensuring data quality		Old machinery cannot supply adequate data		
Lack of internal digital culture and training		Cheap rate labours for mass production		
Resistance to change	Customers do not have the technology	Increasing product variability		
Ineffective change management	Lack of knowledge about providers			
Lack of a digital strategy alongside resource scarcity	Lack of culture collaboration		Multiple customers have different demands	
Strategies for solving the challenges	*Increase automation power to digitise process*	*Retrofit existing machines*		

TABLE B2

Overview of experience of Industry 4.0 implementation of Firm 2

Challenges / Barriers	Horizontal Integration (Upstream)	Vertical Integration	Horizontal Integration (Downstream)	End-to-End Engineering
High Investment in Industry 4.0 Implementation		Cost of implementation runs high until the technology becomes more commonplace		

(Continued)

TABLE B2 (continued)
Overview of experience of Industry 4.0 implementation of Firm 2

Challenges / Barriers	Horizontal Integration (Upstream)	Vertical Integration	Horizontal Integration (Downstream)	End-to-End Engineering
Lack of clarity regarding the economic benefit	Data exchange with suppliers in real-time seems over-engineering		No requirement from the customer	High cost of data storage
Challenge in value chain integration		Training requirement		
Risk of security breaches				
Low maturity level of the desired technology				No data consistency across the lifecycle of products
Inequality				
Disruption to existing jobs	Affected supplier end also as educated experts needed	Skilled experts required that replaces old workforce		High cost of product development
Lack of standards, regulations and forms of certification	Lack of training caused inappropriate data formats	Variance in data formats from departments		
Lack of infrastructure		No server solutions for real-time access to large amounts of data		Data storage facilities are not accessible for SMEs
Lack of digital skills	No digitized solutions	Still some departments require digital skills for data integration	No requirement from the customer	
Challenges in ensuring data quality		Old technologies restricted data collection		
Lack of internal digital culture and training	More focus on operations, lack of investment	Cheaper hand labour as customized products		High cost of product development

Challenges / Barriers	Horizontal Integration (Upstream)	Vertical Integration	Horizontal Integration (Downstream)	End-to-End Engineering
Resistance to change	Lack of knowledge about technology, low investment	Long-serving department heads with little digital experience		
Ineffective change management		Fix department targets		
Lack of a digital strategy alongside resource scarcity	Systems differ as per customers requirement	Different departmental systems requirements	Customer specific orders	
Strategies for solving the challenges	*Stronger digital networking with customers*	*Expansion of digital services with additional services*	*Expert training towards technology adoption*	

TABLE B3

Overview of experience of Industry 4.0 implementation of Firm 3

Challenges / Barriers	Horizontal Integration (Upstream)	Vertical Integration	Horizontal Integration (Downstream)	End-to-End Engineering
High Investment in Industry 4.0 Implementation	Fear of loss of control	Risky investment		High product cost
Lack of clarity regarding the economic benefit				
Challenge in value chain integration		Underdeveloped data analysis		
Risk of security breaches				
The low maturity level of the desired technology				
Inequality				

(Continued)

TABLE B2 (continued)
Overview of experience of Industry 4.0 implementation of Firm 3

Challenges / Barriers	Horizontal Integration (Upstream)	Vertical Integration	Horizontal Integration (Downstream)	End-to-End Engineering
Disruption to existing jobs		Inadequate quality workforce		
Lack of standards, regulations and forms of certification				
Lack of infrastructure		Failure to develop data-based services		
Lack of digital skills		Lack of digital strategy		
Challenges in ensuring data quality				
Lack of internal digital culture and training		Lack of demand for continuous learning		
Resistance to change	Expensive technologies	Lack of senior management support		
Ineffective change management				
Lack of a digital strategy alongside resource scarcity			Multiple customers have different demands	
Strategies for solving the challenges	*Increase automation and upgrading standards*	*Increase networking and digitization, improve current processes*	*Increase automation and upgrading standards*	

14 Industry 4.0 and SME Development

The Case of a Middle Eastern Country

Morteza Ghobakhloo, Masood Fathi, and Parisa Maroufkhani

CONTENTS

14.1 INTRODUCTION

In recent years, the advent of modern digital technologies such as the industrial Internet of Things, additive manufacturing, augmented and virtual reality, and artificial intelligence has given rise to the fourth Industrial Revolution, also known as Industry 4.0 (Müller et al., 2020). As the global economy becomes more digitalized, the manufacturing industry needs to adopt new digital production strategies to keep their competitive positions intact (Müller et al., 2018). Transitioning towards Industry 4.0 represents a massive challenge for small and medium-sized enterprises (SMEs). Industry 4.0 transition is not limited to the mere implementation of advanced digital technologies (Weking et al., 2020). As explained in Figure 14.1, Industry 4.0 and the underlying digital transition is an exceptionally complex phenomenon. Industry 4.0 cannot be regarded as a discrete digitalization project, as it involves developing certain design principles such as vertical and horizontal integration or real-time capability (Ghobakhloo, 2018). Digitalization of industrial value chains and adoption of customer-oriented strategies are among the integral parts of the digital industrial revolution, mandating SMEs to think beyond the internal production processes and

DOI: 10.1201/9781003165880-14

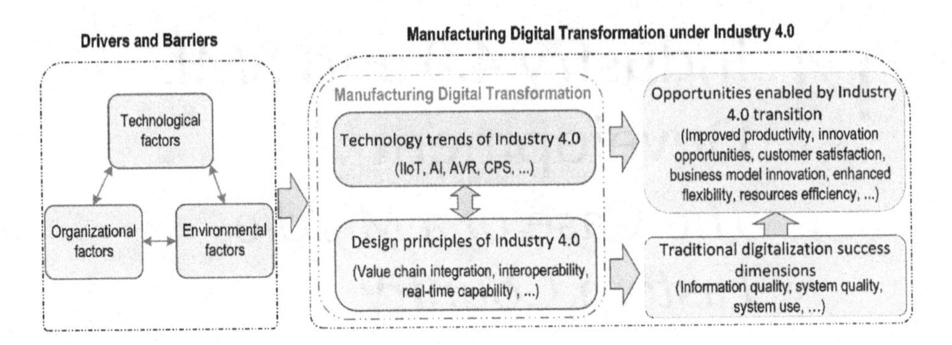

FIGURE 14.1 Industry 4.0 ecosystem.

even supply chain partners when it comes to Industry 4.0 (Ghobakhloo and Fathi, 2019; Masood and Sonntag, 2020).

The disruptive force of Industry 4.0 and the seismic market turbulence intensified by unforeseeable global challenges such as the COVID-19 pandemic is forcing the manufacturing industry to rush towards customer orientation, globalization, and digital integration (Papadopoulos et al., 2020). These radical changes push manufacturers of any size to avoid lagging behind the digitalization race and evade the risk of losing competitiveness (Mittal et al., 2020; Sahi et al., 2020). SMEs are characterized by higher innovation capability and process flexibility, placing them in a superior position in Industry 4.0 (Moeuf et al., 2020). Alternatively, SMEs struggle with resources and expertise that modern digital technologies demanded and fall short in systematically assessing the potential benefits and risks associated with Industry 4.0 transition (Horváth and Szabó, 2019).

Middle Eastern countries are striving to make the most of the digitalization race. The industrial sector in the Middle East is nowadays facing fierce competition and market turbulence. Middle Eastern companies consistently regard the adoption of digital technologies of Industry 4.0, such as cloud computing, artificial intelligence, and additive manufacturing, as a vital strategy for maintaining competitiveness. The present study reports Industry 4.0 and SME development in Iran, one of the biggest and most diverse Middle Eastern economies. Like many Asian economies, SMEs, the manufacturing variants, in particular, are the backbone of Iran's economy (Maroufkhani et al., 2020), making up more than 90 percent of the businesses across the country. SMEs also account for more than 50 percent of national employment and a significant gross domestic product share. The unique economic circumstances in Iran have positioned manufacturing SMEs in a curious position. The recent conflicts on the oil exports, complemented by the devastating COVID-19 crisis, have placed Iranian manufacturing SMEs as vital sources of economic prosperity and growth in the country. Given the essential role of Iranian manufacturing SMEs, the Industry 4.0 transition success is a strategic priority for these businesses. Unfortunately, the Industry 4.0 phenomenon and its application within the Iran SME sector are significantly understudied. Therefore, a detailed investigation of the Industry 4.0 transition phenomenon, in terms of Industry 4.0 drivers, barriers, and opportunities, complemented by a thorough assessment of Iranian SMEs' digitalization level, is highly relevant and needed. Such investigation and the resulting understanding is expected to enable SMEs to develop strategic roadmaps to sense the risks and opportunities of digitalization better and position themselves against Industry 4.0 accordingly.

14.2 COUNTRY BACKGROUND

With an estimated gross domestic product (GDP) of US$610 billion in 2020, Iran is to be ranked as the 21st biggest economy in the world (World Bank Group, 2020). The oil and gas industry has been the backbone of Iran's economy during the past few decades (Morady, 2020). Iran ranks second and fourth in the world for proven natural gas and crude oil reserves, respectively (The World Bank, 2020). Nonetheless, Iran's economy is relatively diversified, with service, agricultural, and manufacturing sectors playing a vital role. Iran is experiencing a 3.5 percent contraction in GPD for 2020–2021, which, compared to many other countries hit hard by the COVID-19 pandemic, is a mediocre contraction. Iran's national development strategy follows a series of predetermined 5-year development plans, which generally value progress in science and technology and economic prosperity and independence (World Bank Group, 2020). Across the 5-year development plans, oil revenue independence has always been a critical goal. Nonetheless, oil revenues are still the primary source of income for the government, which has been volatile during the past two years (The World Bank, 2020).

The recession of the oil and heavy industries and devastating blows of the COVID-19 pandemic to service industries have turned the manufacturing SME sector into the backbone of the country's economy. Over 43,000 SMEs currently operate in various industrial sectors, which account for more than 50 percent of industrial employment in Iran (financialtribune.com). Around 20 percent of Iranian SMEs are struggling with recession, mainly due to the country's unstable economic conditions and the COVID-19 crisis. Nevertheless, manufacturing SMEs active in food, petrochemical, electronic, and textile industries have recorded an over 40 percent increase in international export net value. SMEs have always been a strategic priority for the national and economic development of the country. Various supportive policies and financial incentives have been devised and executed to promote the SME sector, mostly in terms of loans, tax incentives, and consultation services. Even amid revenue crises, the Iranian government runs a comprehensive SME revival program, offering generous financial incentives to reactivate SMEs crippled by the economic and COVID-19 crisis.

14.3 DIGITALIZATION POLICY IN THE COUNTRY

The overall digitalization policy of the country should be viewed in light of Iran's unique sociopolitical norms. Becoming a digitalized economy has been the priority of the various governments during the past two decades, which follows the progress in the science and technology pillar of the country's five-year development plans. Iran is undergoing a digital revolution, placing the country among the fastest-growing digital societies in the world. Although the country witnessed the launch of the first 3G network in 2012, 5G infrastructure and services are rapidly spreading. Out of close to 83 million people, Iran recorded 58.42 million internet users in January 2020, translating to an internet penetration of 70 percent. Compared to 2019, internet users in Iran increased by almost 11 percent. The social media penetration in Iran was recorded 40 percent in 2020, showing a whopping 39-percent year-to-year increase. Although private, social, and business sectors desire digitalization excessively, conservative parties in Iran aim for more restrictive digitalization policies due to cybersecurity considerations. To these conservative parties, digitalization and freedom of information are a double-edged sword, sometimes acting as an existential threat.

As with many other countries, Industry 4.0 is a buzzword within Iran's digitalization policy. The government is slowly adopting supportive Industry 4.0 policies to promote digitalization across various industrial sectors, service, agriculture, and manufacturing sectors in particular. Industry 4.0-–supportive policies of the Iranian government mostly involve promoting academia-industry collaboration for digitalization and increased productivity and efficiency. This initiative also includes funding for pilot research centers that offer a physical space for businesses, practitioners, and academicians to collaborate on developing and implementing Industry 4.0 technologies. The digitalization policy also requires the Iran Ministry of Science, Research, and Technology to force the technical universities to devise new action plans to ensure that academic researchers make a real contribution to the digital transformation of industry as a part of their job requirements. Contrary to the previous digitalization policies implemented in the country, the current digitalization policy, including the Industry 4.0 initiatives, does not include any direct means of financial support for the private sector. The shift in digitalization policies' properties is somewhat attributable to the reduction in the government's revenue since 2018.

14.4 METHODOLOGY

The data used in the present study comes from nationwide research conducted in early 2020, through collaboration with the Ministry of Industry, Mine, and Trade and the Iran Small Industries and Industrial Parks Organization (ISIPO). The nationwide research was conducted to identify the digitalization progress of Iranian manufacturing SMEs under the Industry 4.0 scenario. There is no unified definition of SMEs in Iran, given that various governmental agencies have their particular classification. Nonetheless, the most commonly accepted definition of SMEs in Iran is the following classification proposed by the Iran Central Bank and the Statistical Center of Iran, presented in Table 14.1. This classification is limited to the number of employees and disregards other metrics, such as industry type and revenue. Consistently, nationwide research merely surveyed manufacturing enterprises with full-time employees of fewer than 100 people.

The SME literature widely acknowledges the undeniable direct role of the chief executive officer (CEO) in smaller firms' digitalization efforts as the primary decision maker, profitability analyst, and strategist (Zor et al., 2019). Therefore, CEOs of participating SMEs were targeted as key respondents. The survey and all the

TABLE 14.1
Classification of SMEs in Iran

Enterprise classification	Number of Employees
Microenterprise	Less than 10
Small enterprise	10 to 49
Medium enterprise	50 to 99

preparatory measures, such as the pilot study, were conducted electronically. Overall, 617 Iranian SMEs participated in the electronic survey.

The electronic survey instrument included 41 questions, categorized into four different sections. The first section included 16 items asking about Industry 4.0 barriers experienced by SMEs during various digital transformation stages. These questions employed a 7-point Likert scale, ranging from (1) extremely disagree to (7) extremely agree. The second section consisted of 12 items inquiring into the reasons that drive SMEs to implement digital technologies of Industry 4.0. The 7-point Likert scale, ranging from (1) extremely disagree to (7) extremely agree, was also used for this section. The introductory part of section three asked respondents to recall their perception of digital technologies' potential benefits before the commencement of implementation processes, which later persuaded them to decide on the implementation. Section three included 12 items asking about the extent to which digital technologies implemented have benefited SMEs. Questions in section 4 employed a 7-point Likert scale, ranging from (1) not at all to (7) extremely. The introductory part of section four asked respondents to explain the extent to which digital technologies implemented have benefited them in practice. Section four of the questionnaire measured the implementation rate of various digital technologies.

Table 14.2 explains the rate at which participating SMEs have implemented the digital technologies of Industry 4.0. This table indicates that the overall digitalization rate of Iranian manufacturing SMEs is worryingly low. In particular, the implementation rate of 7 percent or less for the most common technology trends of Industry 4.0, such as artificial intelligence, additive manufacturing, big data analytics, manufacturing simulation, predictive analytics, and virtual reality reveals that Iranian SMEs are significantly behind the digitalization race. Interestingly, not a single participating SME indicated the implementation of blockchain technology. The implementation rate for older and less novel digital technologies such as automated guided vehicles, enterprise resource planning, and machine and process controllers was mediocre, ranging between 10 percent and 30 percent. Surprisingly, 49.27 percent of SMEs had some industrial actuators or sensors implemented, making it the most frequently implemented digital technology among Iranian manufacturing SMEs.

Table 14.2 lists the implementation rate of various digital technologies for medium and small enterprises separately, enabling the direct comparison between them. Results indicate the implementation rate is consistently lower for smaller manufacturers across the majority of digital technologies. The most significant relative discrepancy observed belongs to artificial intelligence, where the implementation rate

TABLE 14.2
Digitalization Rate of Participating SMEs

Digital technology	A total pool of 617 SMEs		A total pool of 332 small enterprises		A total pool of 285 medium enterprises	
	Number of implementors	Implementation frequency	Number of implementors	Implementation frequency	Number of implementors	Implementation frequency
Artificial intelligence	41	6.65%	8	2.41%	33	11.58%
Additive manufacturing	43	6.97%	25	7.53%	18	6.32%
Augmented reality	64	10.37%	23	6.93%	41	14.39%
Automated Guided Vehicles (AGVs) (e.g., Tugger AGVs or outrigger AGV)	203	32.90%	97	29.22%	106	37.19%
Autonomous robots	147	23.82%	51	15.36%	96	33.68%
Big Data analytics	32	5.19%	11	3.31%	21	7.37%
Blockchain technology (e.g., for secure supply chain communications)	0	0.00%	0	0.00%	0	0.00%
Cloud data and storage	80	12.97%	49	14.76%	31	10.88%
Cybersecurity technologies (e.g., Virtual Dispersive Networking)	73	11.83%	28	8.43%	45	15.79%
Enterprise resource planning	188	30.47%	96	28.92%	92	32.28%
High-performance computing-powered computer-aided design	126	20.42%	61	18.37%	65	22.81%
Industrial actuators and sensors	304	49.27%	157	47.29%	147	51.58%
Machine and process controllers (e.g., Distributed control systems or programmable logic controller)	137	22.20%	70	21.08%	67	23.51%
Manufacturing execution system	184	29.82%	95	28.61%	89	31.23%
Manufacturing simulation	33	5.35%	14	4.22%	19	6.67%
Predictive analytics (e.g., data mining, predictive modeling or and machine learning)	44	7.13%	18	5.42%	26	9.12%
Smart industrial wearables/gadgets	184	29.82%	101	30.42%	83	29.12%
Virtual reality	18	2.92%	8	2.41%	10	3.51%

among medium enterprises (11.58 percent) was more than four times higher than the adoption rate of the smaller ones (2.41 percent). Similarly, medium enterprises have enjoyed an implementation rate of more than double in augmented reality, autonomous robots, and big data analytics compared to smaller enterprises. Cloud data and storage is the only digital technology for which smaller manufacturers had a significantly higher implementation rate. This discrepancy is expected, given that cloud-based technologies such as cloud ERP entail lower operating costs such as licensing, software development, or maintenance, and thus better align with resource constraints of smaller businesses.

14.5 INDUSTRY 4.0 BARRIERS

Participating SMEs were asked to highlight how the 16 barriers listed in Table 14.3 have limited their digital transformation efforts. Interestingly, all the 16 barriers listed in Table 14.3 received a mean score of more than 3.5, indicating the significance of each barrier regarding the low rate of digital technology implementation

TABLE 14.3
Industry 4.0 Barriers and Their Significance As Perceived by Iranian Manufacturing SMEs

Industry 4.0 barriers	Mean	Standard deviation
Lack of capital to finance digitalization efforts (technology implementation, training, consultation, etc.)	6.212	1.111
Lack of necessary expertise for the implementation of digital technologies	5.811	1.420
Difficulty to integrate new digital technologies with the existing operations technologies and information technology infrastructure	5.808	1.162
Lack of business partners' willingness to cooperate on digitalization processes	5.733	1.111
Underdeveloped regulation for data security, ownership, and protection	5.723	1.232
Lack of a strategic plan for navigating digitalization efforts	5.674	1.187
Lack of a unified communication protocol across digital assets	5.672	0.889
Lack of managerial competencies to manage change inherent to digitalization	5.403	1.289
Improper and immature organizational structure and procedures for supporting digitalization	5.384	1.027
Lack of managerial competencies to support and streamline digitalization processes	5.160	1.027
Inability to assess the digitalization maturity level before committing to the implementation of digital technologies	5.091	1.052

(*Continued*)

TABLE 14.3 (continued)
Industry 4.0 Barriers and Their Significance As Perceived by Iranian Manufacturing SMEs

Industry 4.0 barriers	Mean	Standard deviation
Cybersecurity security concerns	4.942	1.446
Lack of employee readiness for implementation of digital technologies	4.865	1.312
Unfamiliarity with potential benefits that digital technologies might offer	4.720	1.066
Employee resistance to change inherent to the implementation and utilization of digital technologies	4.123	0.913
Unfamiliarity with potential risks that might be associated with digital transformation	3.889	1.087

within Iran's manufacturing SME sector. Findings show that lack of capital to finance digitalization efforts is the most significant barrier towards digitalization. This finding is not particularly unexpected, given the lack of financial resources has long been acknowledged as the primary barrier to technology implementation within the SME literature (AlBar and Hoque, 2019). Prior studies frequently highlighted the concern regarding Iranian SMEs struggling with financing digitalization projects, and the present survey revealed that lack of capital is still the dominating barrier for Iranian SMEs, even within the Industry 4.0 era. Lack of necessary expertise for implementing digital technologies ranks as the second most crucial barrier within Iranian SMEs. The SME digitalization literature widely acknowledges the overall lack of information and digital technology knowledge and expertise within smaller organizations (Ghobakhloo and Ng, 2019) and recognizes it as one of the major obstacles towards digital transformation among SMEs (Horváth and Szabó, 2019; Kumar et al., 2020).

Difficulty in integrating new digital technologies with the existing information technology (IT) and operations technology (OT) infrastructure is the third most significant barrier towards digitalization, as perceived by Iranian SMEs. This finding is not surprising, given digital transformation under Industry 4.0 requires organizations to have the necessary digitalization and OT maturity to develop an interoperable manufacturing ecosystem where layers of systems and technologies can communicate with each other and share data in real time. The industrial and academic report indicates that SMEs generally lack the necessary IT and OT maturity when it comes to digital transformation (Stentoft et al., 2020). Table 14.3 also reveals that the lack of business partners' willingness to cooperate on digitalization processes ranks as the fourth major barrier towards digitalization. Industry 4.0 mandates Iranian SMEs to deal with the hyper-competition and reduced time to market. Therefore, the survival of Iranian manufacturing SMEs today depends on their ability to collaborate with the business partners to form mutual digitalization efforts and policies to create digitalized supply networks. Results also indicate that underdeveloped regulation for data security, ownership, and protection is a major barrier limiting Iranian SMEs' digitalization efforts. This finding suggests the inefficiency of existing general data regulation policies adopted by the government.

A closer look at Table 14.3 reveals that a few significant barriers of Industry 4.0 concern Iranian SMEs' incompetency for the strategic management of digital transformation efforts. This cluster of management-related obstacles includes lack of a strategic plan for navigating digitalization efforts, lack of a unified communication protocol across digital assets, lack of managerial competencies to manage change inherent to digitalization, and lack of managerial competencies to support and streamline digitalization processes. This finding highlights the importance of SMEs' strategic competency in effectively managing digitalization processes. Though they are less often emphasized, cybersecurity concerns and employees' resistance to change are other significant barriers to Industry 4.0 as perceived by Iranian manufacturing SMEs. Finally, the potential risk of digitalization was regarded as the least considerable barrier by participating SMEs.

14.6 INDUSTRY 4.0 DRIVERS

Participating SMEs rated the 12 drivers of Industry 4.0 as shown in Table 14.4. Having greater flexibility to better react to market changes was the most significant driver for implementing the digital technologies of Industry 4.0 among Iranian manufacturing SMEs. Staying competitive in the hyper-competitive market and improving the overall quality of products and services were, respectively, the second and third most important reasons for digitalization as perceived by the participating

TABLE 14.4
Industry 4.0 Drivers and Their Significance as Perceived by Iranian Manufacturing SMEs

Industry 4.0 drivers	Mean	Standard deviation
To increase flexibility to better react to changes in the market	5.407	1.164
To stay competitive in the hyper-competitive market	5.329	1.253
To improve the overall quality of products and services	5.006	1.251
To decrease the costs of operation	4.861	0.927
To satisfy customer and business partner requirements	4.658	0.977
To fulfill the requirement for joining the international supply chains	4.573	0.983
To improve the process, operation, and performance monitoring and control	4.457	1.229
To contribute to the strategic goals and values of the organization	4.416	1.016
To improve the productivity of employees	4.385	1.241
To increase the overall profitability of the organization	4.370	1.039
To reduce the time-to-market of products and services	3.842	0.988
To improve the corporate image	3.255	1.309

SMEs. It is an exciting result, revealing that Iranian SMEs face hyper-competition, and they see digitalization as a strategy to deal with the ever-increasing competitive pressure. This finding is further supported by participant perception of the two popular drivers of employee productivity improvement and profitability enhancement, placing them among the less important drivers of Industry 4.0. Cost efficiency, customer satisfaction, internationalization, better monitoring and control, and strategic value creation are among the more intermediate drivers of Industry 4.0 as perceived by participants. Table 14.4 shows that out of 12 drivers, ten drivers have the mean value of 4.3 or higher, with improving the corporate image as the only driver with the mean value of less than 3.5. This finding implies that Iranian SMEs regard digitalization as a strategic game changer.

14.7 INDUSTRY 4.0 OPPORTUNITIES

Previous industrial revolutions have not been too kind to Iranian SMEs. Smaller manufacturers in Iran are significantly behind the larger manufacturers in terms of technological advancement. Overall, many pitfalls and barriers have prevented Iranian SMEs from reaching the necessary economies of scale and scope. Nonetheless, with the rapid development of information and communications technology infrastructure within the country and more accessible and commercially available digital technologies, particularly digital products developed domestically, Iranian manufacturing SMEs now have the opportunity to rely less on economies of scale and scope for business growth and competitiveness.

Industry 4.0 technologies such as cloud services now allow Iranian SMEs to operate in the data-driven economy, even if only at the local scale. Iranian manufacturers now have the opportunity to collect the market data, analyze it purposefully, and have a deeper understanding of customer preferences and market trends. This capability allows Iranian manufacturing SMEs to compete against larger counterparts in market share at the local and even regional levels. Iranian SMEs can draw on Industry 4.0 technologies to better understand consumer behavior and future preferences and incrementally alter their product or service portfolios accordingly. Such a digitalization and integrability level would allow Iranian SMEs to gradually develop the necessary competencies for entering into the data-driven international value chains.

Productivity improvement is another opportunity driving the progress of Industry 4.0 and digitalization in Iran. The majority of Iranian manufacturing SMEs struggle with productivity excellence, and Industry 4.0 offers significant opportunities to Iranian firms to address the internal productivity concerns. Industry 4.0 can enable Iranian manufacturers to enjoy the benefits of technology leapfrogging by accessing the most advanced and novel digital and operations technologies that are today mature, commercially available, and financially accessible, even to smaller firms. In reality, Industry 4.0 potential for technological leapfrogging offers Iranian SMEs the opportunity to learn from industry leaders' technological development experience and avoid the risk and costs of implementing the successive generations of digital manufacturing technologies.

Business resilience is arguably the most crucial opportunity that drives Industry 4.0 among Iranian industrial sectors. Due to well-known political conflicts and the COVID-19 crisis, Iran faced the largest economic crisis in recent history. During the past 18 months, Iranian SMEs across various sectors faced many difficulties maintaining operations and protecting employees. Since the international conflicts and the COVID-19 crisis disrupted many avenues of international trade for importing necessary goods, most Iranian SMEs had to deal with the shortage of raw materials or massive spikes in demand and market preferences. Since flexibility is a core principle of Industry 4.0, the digital transformation under Industry 4.0 can enable Iranian SMEs to increase the flexibility of their operations, better react to changes in the business environment, and survive the ongoing crisis.

14.8 CONCLUSION

Industry 4.0, fueled by the application of advanced digital technologies, is transforming the business market and rules of competition. Like many other countries, Industry 4.0 and the implementation of advanced digital technologies within the SME manufacturing sector are believed to provide Iran with enormous economic growth opportunities. Nonetheless, the statistics reveal that Iranian SMEs are facing major challenges in the wake of digital transformation. The low adoption rate observed across a wide range of digital technologies, particularly among small manufacturers, is strong evidence of these challenges. Observing such challenges within a developing economy is expected, given SMEs in industrial economies such as Italy, Germany, and Sweden, as leaders of Industry 4.0 transition, struggle with more or less similar challenges.

Overall, SMEs are in an inferior position when it comes to digitalization under Industry 4.0, mainly because of the resource constraints inherent to their nature. In response to challenges faced by SMEs, governments worldwide are adopting supportive digitalization policies that prioritize SMEs' Industry 4.0 transition. Although the supportive policies across various governments differ in design and incentivizing approaches, they are usually unified in their core objectives. Providing some financial incentives has been an indispensable part of any world-class Industry 4.0 supportive policy. Unfortunately, Iran's economy has been crumbling in the face of the pandemic crisis. Contrary to the preceding national digitalization policies that always involved a form of financial incentives, the Iranian government nowadays does not provide any explicit financial aid to support digitalization directly, not even in the form of tax incentives. Not surprisingly, the lack of financial resources to fund the digitalization efforts such as technology implementation or employee training is the most significant barrier towards the Industry 4.0 transition within Iran's SME sector. Overall, Iranian SMEs lack the capability to develop an all-inclusive strategic plan for navigating digitalization efforts and correctly set their digitalization objectives. Although Iranian SMEs have benefited, partially, from some opportunities of Industry 4.0, nonetheless, there is a significant difference between what Iranian SMEs expect of digital transformation and what they have in reality achieved as a result of implementing digital technologies of Industry 4.0. This discrepancy has rendered the competitive advantages of Industry 4.0 for Iranian SMEs somewhat accidental.

REFERENCES

AlBar, A. M., & Hoque, M. R. (2019). Factors affecting the adoption of information and communication technology in small and medium enterprises: A perspective from rural Saudi Arabia. *Information Technology for Development*, *25*(4), 715–738.

Bank, T. W. (2020). Islamic Republic of Iran. Retrieved from www.worldbank.org/en/country/iran/overview.

Ghobakhloo, M. (2018). The future of manufacturing industry: A strategic roadmap toward Industry 4.0. *Journal of Manufacturing Technology Management*, *29*(6), 910–936.

Ghobakhloo, M., & Fathi, M. (2019). Corporate survival in Industry 4.0 era: The enabling role of lean-digitized manufacturing. *Journal of Manufacturing Technology Management*, *31*(1), 1–30.

Ghobakhloo, M., & Ng, T. C. (2019). Adoption of digital technologies of smart manufacturing in SMEs. *Journal of Industrial Information Integration*, *16*, 100107.

Group, W. B. (2020). Iran economic monitor: Mitigation and adaptation to sanctions and the pandemic, Spring 2020. Retrieved from http://documents1.worldbank.org/curated/en/229771594197827717/pdf/Iran-Economic-Monitor-Mitigation-and-Adaptation-to-Sanctions-and-the-Pandemic.pdf.

Horváth, D., & Szabó, R. Z. (2019). Driving forces and barriers of Industry 4.0: Do multinational and small and medium-sized companies have equal opportunities? *Technological Forecasting and Social Change*, *146*, 119–132.

Kumar, R., Singh, R. K., & Dwivedi, Y. K. (2020). Application of industry 4.0 technologies in SMEs for ethical and sustainable operations: Analysis of challenges. *Journal of Cleaner Production*, *275*, 124063.

Maroufkhani, P., Ismail, W. K. W., & Ghobakhloo, M. (2020). Big data analytics adoption model for small and medium enterprises. *Journal of Science and Technology Policy Management*, ahead-of-print.

Masood, T., & Sonntag, P. (2020). Industry 4.0: Adoption challenges and benefits for SMEs. *Computers in Industry*, *121*, 103261.

Mittal, S., Khan, M. A., Purohit, J. K., Menon, K., Romero, D., & Wuest, T. (2020). A smart manufacturing adoption framework for SMEs. *International Journal of Production Research*, *58*(5), 1555–1573.

Moeuf, A., Lamouri, S., Pellerin, R., Tamayo-Giraldo, S., Tobon-Valencia, E., & Eburdy, R. (2020). Identification of critical success factors, risks and opportunities of Industry 4.0 in SMEs. *International Journal of Production Research*, *58*(5), 1384–1400.

Morady, F. (2020). *Contemporary Iran: Politics, Economy, Religion*. Bristol: Bristol University Press.

Müller, J. M., Buliga, O., & Voigt, K.-I. (2018). Fortune favors the prepared: How SMEs approach business model innovations in Industry 4.0. *Technological Forecasting and Social Change*, *132*, 2–17.

Müller, J. M., Buliga, O., & Voigt, K.-I. (2020). The role of absorptive capacity and innovation strategy in the design of industry 4.0 business Models-A comparison between SMEs and large enterprises. *European Management Journal*, *39*(3), 333–343.

Papadopoulos, T., Baltas, K. N., & Balta, M. E. (2020). The use of digital technologies by small and medium enterprises during COVID-19: Implications for theory and practice. *International Journal of Information Management*, 102192.

Sahi, G. K., Gupta, M. C., & Cheng, T. (2020). The effects of strategic orientation on operational ambidexterity: A study of indian SMEs in the industry 4.0 era. *International Journal of Production Economics*, *220*, 107395.

Stentoft, J., Adsbøll Wickstrøm, K., Philipsen, K., & Haug, A. (2020). Drivers and barriers for Industry 4.0 readiness and practice: Empirical evidence from small and medium-sized manufacturers. *Production Planning & Control*, 1–18.

Weking, J., Stöcker, M., Kowalkiewicz, M., Böhm, M., & Krcmar, H. (2020). Leveraging industry 4.0 – A business model pattern framework. *International Journal of Production Economics*, *225*, 107588.

Zor, U., Linder, S., & Endenich, C. (2019). CEO characteristics and budgeting practices in emerging market SMEs. *Journal of Small Business Management*, *57*(2), 658–678.

15 Industry 4.0 in Russia
Digital Transformation of Economic Sectors

Pavel Anatolyevich Drogovoz, Nataliya Aleksandrovna Kashevarova, Vladimir Alekseevich Dadonov, Tatyana Georgievna Sadovskaya, and Maksim Konstantinovich Trusevich

CONTENTS

15.1 INTRODUCTION AND COUNTRY BACKGROUND

The land area of the Russian Federation measures more than 17 million km2 (The Central Intelligence Agency, 2020); the population was estimated at 146,748,590 as of the beginning of 2020 (The Central Intelligence Agency, 2020; Federal State Statistics Service, 2020). In 2018, the nominal GDP was $1,657 billion and GDP per capita measured $28,797; share of industry in GDP was 32.4 percent and share of high-technology and science-intensive economy sectors was 21.1 percent combined (Federal State Statistics Service, 2020; International Monetary Fund, 2020). Extractive industries comprised of oil, natural gas, metal, and other minerals production constituted 46.1 percent of gross value added in 2018 (Federal State Statistics Service, 2020). Export volume in 2017 measured $353 billion with major exports being oil, oil products, natural gas, metals, a wide range of civilian and military goods, and petrochemical and timber industries' produce (The Central Intelligence Agency, 2020). Based on 2017 data, Russia's key export partners are China (10.9 percent), the Netherlands (10 percent), Germany (7.1 percent), Belarus (5.1 percent), and Turkey (4.9 percent) (The Central Intelligence Agency, 2020).

DOI: 10.1201/9781003165880-15

The Fuel and Energy Complex (FEC) accounts for about half of Russia's export revenues generating a significant share of federal budget income, but the same time, the share of non-primary exports tends to grow. The Russian FEC consists of natural gas, oil, coal, and nuclear industries, as well as renewable energy sources. The military-industrial complex includes a large amount of scientific organizations and, due to intensive diversification processes, supports high-technology developments in related industries. The aerospace industry, being a highly significant entity in the modern world, is another driver of innovation.

The public sector takes up a large part of the Russian economy. In order to reform public governance in a number of industrial sectors, consolidate enterprises, and support civilian, military, and mixed high-technology manufacturing, a few public corporations were established in the mid-2000s. Some of the most notable of these are as follows:

- Rostec: a public corporation supporting development, manufacturing, and export of high-technology industrial products.
- Rosatom: a public corporation focused on nuclear energy.
- Roscosmos: a public corporation focused on space activities.

15.2 DIGITALIZATION POLICY IN THE COUNTRY

In 2018, the Ministry of Digital Development, Communications, and Mass Media was created by a presidential decree under the auspices of the Ministry of Communications and Mass Media. The new ministry's goals are:

- Facilitation of mostly digitalized provision of public, municipal, and socially significant services.
- Facilitation of rapid development of information technology industry.
- Provision of equal access to communication services, the internet, and media environment.

The ministry oversees the national program, "Digital Economy." The program was adopted in 2018 as well and is expected to last until 2024. The program is aimed at:

- Formation of a new legislative environment for relations between citizens, businesses, and the state arising from the development of digital economy.
- Creation of a modern and secure high-bandwidth infrastructure for storage, processing, and transfer of data.
- Creation of a personnel training system for the digital economy.
- Facilitation of projects aimed at implementation of promising digital technologies.
- Increasing public governance efficiency and effectiveness via introduction of digital technologies and industrial solutions.

(Ministry of Digital Development, Communications and Mass Media of the Russian Federation, 2020a)

The national program's target is to increase the share of domestic digital economy development expenses in GDP from 1.7 percent in 2017 to 5.1 percent in 2024. The gross budget of the program amounts to ₽1634.9 billion. The national program includes the following projects:

- Regulation of digital environment (₽1.7 billion).
- Information infrastructure (₽772.4 billion).
- Personnel for the digital economy (₽143.1 billion).
- Cybersecurity (₽30.2 billion).
- Digital technologies (₽451.8 billion).
- Digital public governance (₽235.7 billion).

In addition, an autonomous non-profit organization, also called "Digital Economy" was created in order to support projects in the field of digital economy and coordinate cooperation between business community, research organizations, and public authorities; the organization was co-founded by the Russian government and a number of public and private companies. With the support from this organization, roadmaps for development of end-to-end digital technologies were designed. The national program "Digital Economy" foresees the development of a number of technological operations combined into end-to-end digital technologies. They are shown in Table 15.1.

TABLE 15.1
List of Subtechnologies within End-to-End Digital Technologies

End-to-end digital technologies	Subtechnologies
Neurotechnology and Artificial Intelligence (AI)	– Computer vision – Natural language processing – Speech recognition and synthesis – Recommendation systems and intelligent acceptance support systems – Advanced methods and technologies in AI – Neuroprosthetics – Neurointerfaces, neurostimulation and neurosensing
Robotics Components and Sensing	– RTK sensors and digital components for human-machine interaction – Sensorimotor coordination and spatial positioning technologies – Sensors and sensory processing
Distributed ledger systems	– Technologies for ensuring data integrity and consistency (consensus) – Technologies for creating and executing decentralized applications and smart contracts – Data organization and synchronization technologies

(*Continued*)

TABLE 15.1 (continued)
List of Subtechnologies within End-to-End Digital Technologies

End-to-end digital technologies	Subtechnologies
Wireless technologies	– WAN (Wide Area Network) – LPWAN (Low Power Wide Area Network) – WLAN (Wireless Local Area Network) – PAN (Personal Area Network) – Satellite communication technologies
New production technologies	– Digital design, mathematical modeling and product or the whole production output lifecycle management (Smart Design) – Smart Manufacturing – Manipulators and manipulation technologies
Quantum technology	– Quantum computing – Quantum communications – Quantum sensors and metrology
Virtual and augmented reality technologies	– Development tools for VR/AR content and technologies for improving user experience (UX) from the side of developer – Platform solutions for users: content creation and distribution editors – Motion capture technologies in VR/AR and photogrammetry – Feedback interfaces and sensors for VR/AR – Graphics output technologies – Data transfer optimization technologies for VR/AR

Another entity should be mentioned as well: the National Technological Initiative (NTI) is an association bringing together business and expert communities' representatives in order to support development of promising technology markets and industries in Russia (National Technology Initiative, 2020). As part of the initiative, a roadmap, called "Technet," was designed with its aim being the development and implementation of cutting-edge manufacturing technologies within the timeframe of 2017–2035.

One vivid example of the national program's implementation of a digital economy at the sectoral level is the nuclear industry. The Rosatom public corporation has developed a unified digital strategy aimed at the acceleration of a sector's digital transformation. The priority areas of the strategy are digitalization of the corporation's internal processes and functions, development and market placement of Rosatom's digital products, and participation of the corporation in the development of digital economy. In addition, the strategy envisages cooperation with other participants of the digitalization process – both domestic and foreign.

Another active participant in implementation of the national Digital Economy program is the Rostec public corporation, which became a center for "development of research competences and laying the technological groundwork" within five technological areas: neural networks and artificial intelligence, industrial internet, robotics and sensor components, wireless connection technologies, and

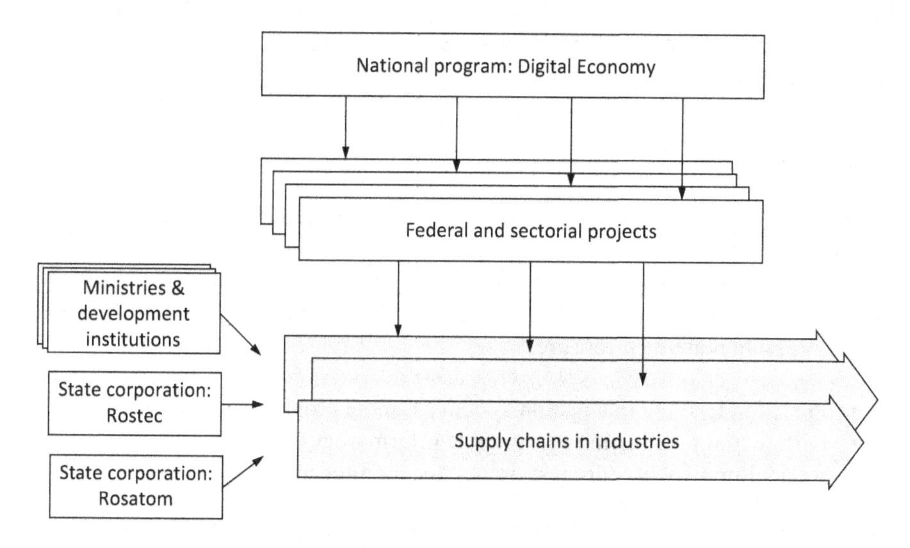

FIGURE 15.1 Digital economy development policy in Russia.

distributed ledger systems. Being responsible for the technological groundwork, Rostec searches for promising projects and determines the technological leaders, whose collaboration with Rostec opens up new opportunities for digital systems' development.

The Ministry of Agriculture of the Russian Federation has developed the Digital Agriculture project, which will last until 2024. The goal of the project is the digital transformation of agriculture through the introduction of digital technologies and platform solutions to ensure technological breakthroughs and productivity growth in digital agricultural enterprises. The state corporation, Rostec, is participating in the development of solutions for this project. Rostec's technology portfolio includes software systems for farm management, robotic systems, drones for monitoring agricultural facilities, and precision farming technologies based on the Internet of Things (IoT; Rostec State Corporation, 2020).

The implementation of the digital economy development policy is summarized in Figure 15.1.

15.3 METHODOLOGY

The following methods were used in this study:

- Review of publications in business media.
- Interviews with experts involved in Industry 4.0 projects from public and private organizations.
- Business case analysis.

The sample of business cases was formed by selecting a number of Small and medium enterprises (SMEs) that are (1) in the process of digital transformation,

(2) belong to the manufacturing industry, and (3) whose annual turnover does not exceed €50 million.

15.4 OPPORTUNITIES

In the background of these positive trends, digital technologies enable Russian agricultural enterprises to make a qualitative leap in development and significantly reduce losses throughout the entire life cycle of production. Today, many enterprises in the industry are actively moving in this direction. The most popular technologies among Russian manufacturers are:

- Precision farming (navigation systems, remote probing).
- Differential fertilization (geographic information systems).
- Agricultural robots (drones, drones for monitoring the state of fields and harvesting, smart sensors).
- AIoT platforms (control of data from sensors and equipment, vertical farms).
- Big data (analysis of data received from sensors to make an accurate forecast and strategy).

(Ivanov, A., 2021)

Another promising area is the digitalization of the forest industry. One-fifth of the world's forest cover is located in Russia, forests occupy 46.4 percent of the country's territory, but the efficiency of using Russian forests leaves much to be desired – Russia's share in the global volume of forestry complex (FC) production is only 3 percent, and the added value per unit of harvested timber in Russia is two times lower than in Canada, and six times lower than in Finland (Vilde, O., 2021). In logging, digitalization means the transition to the interaction of forestry machines and equipment to work with big data. This is data from satellites, 3D-LiDAR, and drones about the relief, digital terrain model in logging areas, weather conditions and other external factors. Digital maps and terrain models help market participants assess and update the state of the forest fund, monitor the progress of reforestation, and predict future volumes of haulage and supply of raw materials.

Predictive analytics help predict potential component failures and downtime. Also, systems based on artificial intelligence allow synchronizing the work of forestry machines to increase the efficiency of felling, ensure the laying of optimal routes for transporting wood. Machine vision technologies are expected that will allow loggers to determine the size and quality characteristics of literally every tree.

At the end of 2020, the Russian government began discussing the possibility of creating a network of data centers in the Arctic. It is assumed that private business will be engaged in the construction of data centers, and the state will develop a package of measures to support IT residents of a special economic zone (SEZ) in the Arctic (Skobelev, V., Galimova, N., & Skrynnikova, A., 2020). Low temperatures can significantly reduce the cost of cooling data center equipment, and similar projects are already being implemented in Sweden. The construction of data centers in the cold zone, in addition to the need to cool the equipment, will also solve social problems: surplus heat can be used to heat living quarters and work premises, and the implementation of the project will contribute to the development of telecommunications infrastructure in the settlements of the region. Experts note that the economic model and the feasibility of such a project are affected by problems with guaranteed electricity, infrastructure, communication channels, and qualified personnel. Part of the problems will be solved by a project to lay a fiber optic communication line along the bottom of the Arctic Ocean, which the Russian company "Megafon" began to implement in 2020 together with the Finnish infrastructure operator "Cinia" (Kinyakina, E., 2020). At the end of 2020, the opening of the first data center in Karelia was planned, and the construction of the second was also announced. It is expected that the clients of these data centers will be engineering, metallurgical, and IT companies located in Karelia.

The industries discussed earlier have the potential for exponential growth if digitalization is fast and smart. Due to the specifics of the Russian economy, pilot projects are implemented with the participation of large state-owned or state-controlled corporations. But for small and medium businesses, this shouldn't be a problem because giants are taking pioneer risks and shaping supply chains.

15.5 DRIVERS

According to experts the introduction of digital technologies is one of the key drivers of economic growth, the digitalization of industries leads to a change in the demand for factors of production, and by 2030 the contribution of digitalization to Russia's GDP will be 18.4 percent compared to 1 percent in 2017 (Abdrakhmanova et al., 2019; Gromoff et al. 2014). The total amount of budgetary funding under the federal program, Digital Technologies, is 280 billion rubles until 2024, and in 2019, within the framework of the program, 306 projects were approved for the development and implementation of digital platforms, the development of digital technologies, and the support of leading companies for a total amount of 14 billion rubles (Digital Economy Digest, 2019a). The main trends in the digital transformation of companies in Russia in 2020–2021 are:

- The emergence and development of digital platforms in certain industries and related new opportunities and threats to business.
- Transforming customer experience, accelerating digital adoption, and moving to a data-driven governance model.
- Accelerating the transition to a new generation of digital infrastructure for companies: flexible, open, cloud-based, data integration.
- Overcoming the shortage of employees and competencies for new jobs simultaneously with automation and the need to reduce traditional jobs, adaptation to new work formats and models of competence development (Analytical Center under the Government of the Russian Federation, 2020).

At the same time, there is a large gap in the level of digitalization of companies, even within the same industry. Only about a third of companies are moving from the implementation of individual digital solutions to a wider digital transformation: wider digital transformation that includes the introduction of new digital business models and products, work with digital personnel and culture. The number of leading companies is small, and for them the business model is the driver of digital transformation.

It should be noted that Russia has a strong scientific base in the field of Industry 4.0 technologies. According to a study (Ministry of Digital Development, Communications and Mass Media of the Russian Federation, 2020b) conducted in 2017, Russian applicants have filed in Russia and abroad only 4,276 patent applications for inventions related to end-to-end digital technologies, which is 27 percent more than in 2010. The largest trend towards an increase in the number of applications from 2010 to 2017 was observed for end-to-end digital technologies "New production technologies," "Neurotechnologies and artificial intelligence," and "Technologies of virtual and augmented reality." As shown in the study (Ministry of Digital Development, Communications and Mass Media of the Russian Federation, 2020c), in terms of the number of publications for the end-to-end digital technologies under consideration, Russia ranks in the TOP-25, and as of 2018, Russia's contribution to the total number of publications worldwide is about 2 percent.

Opportunities for financing digital transformation projects for consumer companies and solution providers through government support programs are expanding, which is especially important for small and medium-sized businesses. For example, the state institute for the development of the venture market, JSC Russian Venture Company, organizes competitions for grants for developers of products, services, and platform solutions based on end-to-end digital technologies.

The state corporation Rosatom has a unified sectoral procedure for the implementation of innovative solutions for small and medium-sized businesses. A similar "One Window" system for the introduction of innovations in cooperation with small and medium-sized businesses was created in PJSC Rosneft.

An innovation cluster has been created in Moscow to create the conditions necessary for the effective development of innovations and new projects. The i.moscow digital cluster platform contains, among other things, the following services:

- Navigator of government support measures.
- Access to venture investment.
- Contract manufacturing exchange.
- Factoring.
- Marketplace of products and services.
- Technology competitions.

Also, the Moscow authorities are implementing the smart city concept. Within the framework of this concept, an intelligent transport infrastructure and a video surveillance system are being developed, unified systems for providing public services to the population are being created, access to free city Wi-Fi is expanding, and urban management is being improved. In 2020, Moscow took 56th place in the Smart City Index rating of the Swiss business school IMD (Tadviser, 2020). The proliferation of IoT technologies is playing a significant role in the digital transformation. The volume of the IoT market in Russia in 2019 amounted to $ 3.7 billion, and the average annual growth rate in the period from 2018 to the end of 2023 will be 19.7 percent (Institute for Statistical Research and Knowledge Economics, HSE, 2019).

The concept of construction and development of narrowband wireless communication networks of the IoT (Digital Economy Digest, 2019c) involves the active use of the IoT in the following sectors of the economy:

- Housing and communal services (connected meters for monitoring water and electricity supply, ensuring effective management of fixed assets and infrastructure).
- Transport (traffic management, cargo tracking, monitoring the condition of vehicles and rolling stock).
- Industry (control and maintenance of serviceability of industrial equipment).
- Healthcare (control over patients, medical personnel, equipment, dispensing drugs).
- Agriculture (efficient use of land resources, funds, storage of products).

The digital transformation of industry is closely related to the development of 5G networks. The 5G standard includes communication protocols of the LPWA (Low Power Wide Area) category, which are designed for low-intensity communication of a large number of subscribers with a very low level of power consumption of modems. The 5G slot can be connected to an infinite number of devices continuously transmitting a small amount of data. As of the beginning of 2021, 5G zones in various cities are being launched in pilot mode in Russia. Mostly these are dedicated networks on the territory of industrial enterprises.

Digital twins are a hotly debated technology in mechanical engineering. Modern mechanisms, from mechanical benches to cars, generate a huge amount of information that is accumulated and analyzed by the manufacturer. On the basis of the data obtained, it is possible to correct the operation of the unit or send its time. In Russia, such digital twins are already being created for shunting and mainline locomotives. Based on the collected data, a predictive analysis of the performance of the nodes was carried out (Ivanov, A., 2021). Similar technical solutions are being developed by manufacturers of cars and stationary facilities, primarily utilities infrastructure. Transmitting data with minimal signal latency over 5G networks will allow to manage networks of drones and unmanned vehicles, optimizing their movement around the city and distributing flows.

Also, today there is an increase in requirements for production flexibility, which, in turn, increases the demand for industrial robotics from companies to modernize production. The range of applications for industrial robots is quite wide. These are industrial robotics and additive technologies, self-driving cars, logistics robotics, drones, unmanned vehicles, and mining and construction robotics. In many areas, robots are capable of replacing humans in conditions that are dangerous to life and health. In the technological aspect, the driver for the development of robotics technologies is the increase in the capacity of fuel cells, the development of self-loading sensor devices and image recognition technologies. The use of robotics in modern conditions cannot be imagined without artificial intelligence, which is necessary to control robots. Components such as computer vision and unmanned vehicles are also critical. The introduction of artificial intelligence in industry is also facilitated by the need for companies to increase the speed of business processes while reducing costs and increasing investment in the development of artificial intelligence on the part of companies. In addition, in the past decade, a significant incentive for the widespread penetration of all digital technologies has been the reduction in the cost of technological solutions, such as sensors for the industrial Internet, and information storage.

A group of national standards in the field of smart manufacturing is being developed and is already partially being implemented, including digital twins of production, methods and technologies of mathematical modeling and virtualization of product testing. The development of documents was initiated to form a full-fledged ecosystem of regulatory and technical regulation of the digital industry. They are expected to have a significant impact on the accelerated digitalization of the industrial sector.

Thus, four Industry 4.0 drivers can be distinguished:

1. Technological: the development of digital technologies that shape various aspects of the Industry 4.0 system. This is an increase in the efficiency of various robotics systems, the active development of solutions in the field of artificial intelligence and an increase in the performance of computing systems, the development of energy-efficient long-range communication technologies.
2. Institutional: federal, sectoral, and municipal programs aimed at developing and supporting digital transformation processes.
3. Corporate: cost reduction and increased availability of technological solutions, the need to accelerate business processes while reducing costs, increased investment by companies in the development of Industry 4.0 technologies.
4. Market: increased requirements for production flexibility, increased demand for customized products, the possibility of more effective interaction with partners in the market.

15.6 BARRIERS

Despite a significant number of successful digitalization projects at the industry and corporate levels, the level of digital maturity of most Russian organizations remains rather low, while the level of investment required for the development and implementation of digital technologies is too high for them. Digital business transformation is associated with changing business models (Müller, Buliga, & Voigt, 2020), (Müller & Voigt, 2018). Within the framework of the federal project, Digital Technologies, important aspects are (Abdrakhmanova et al., 2019):

- Development of chains "markets – products – technologies"
- Taking into account not only the positions of development organizations, but also the prospective demand of consumers.

- Taking into account the existing groundwork and competitive Russian end-to-end digital technologies, identifying "gaps" in the end-to-end digital technologies.
- The interconnection of individual roadmaps with each other, especially in terms of close or intersecting end-to-end digital technologies, such as artificial intelligence, big data, new production technologies, robotics, etc.

The roadmaps for the end-to-end digital technologies development provide an assessment of the readiness level (TRL) of these technologies in Russia. So, the level of readiness for the end-to-end digital technology "New production technologies" for certain specific single examples is TRL 6–9, but the results of the expert survey indicate an insufficient number of solutions with high TRL on the market, and the lag in the development of most subtechnologies in Russia is 5–10 years in comparison with the world level (Digital Economy Digest, 2019b). In this case, the following positive factors are observed:

- Democratization of computer engineering technologies.
- The need to reduce time to market.
- Increasing the performance of computing systems.

According to a study carried out by the JSC Central Research Institute, Elektronika, jointly with the Autonomous non-profit organization Digital Economy and the *Applied Informatics* publication, the IoT industry in Russia is still in its early stages of development. The overwhelming majority of respondents interviewed for the study believe that the insufficiently dynamic development of the industry is associated with the absence of a state program to support this segment. In particular, several factors influence the current situation:

- Low level of communication technologies today.
- Insufficient number of certification tests.
- Uncertainty about the effectiveness of the implementation of IoT technologies.

Organizations that position themselves as players in the IoT market are not currently engaged in relevant projects but are trying to occupy a niche in order to further have the status of a full-fledged player. This approach speaks of the prospects of the technology and the interest of enterprises in the implementation of the development of the IoT, but does not allow the market to develop dynamically.

As of the beginning of 2021, the problem with frequencies for 5G networks has not been solved in Russia: the plan is to build 5G networks at frequencies of 4.7–4.9 GHz, while most countries use 3.4–3.8 GHz. In Russia, this range is occupied by military satellite networks, and the 4.7–4.9 GHz range is used in "friend or foe" detection systems in military equipment of NATO countries. In this regard, the International Telecommunication Union issued a requirement according to which countries wishing to build internal communication infrastructure using the 4.7–4.9 GHz band at a distance of less than 300 kilometers from other states must request their permission.

In 2018, only 23.6 percent of organizations in the business sector in Russia used software tools for managing automated production and/or individual technical means and technological processes in their activities (Institute for Statistical Research and Knowledge Economics, HSE, 2020). Figure 15.2 shows the distribution by industry.

According to World Robotics, in 2019 Russia showed a low level of robot penetration – 5 robots per 10 thousand people, which is 20 times lower than the world average. As for the levels of readiness of subtechnologies, the sensors and digital components are at the level of TRL-7, technologies of sensory-motor coordination – TRL-6, the same level for sensors and processing of sensory information (Digital Economy Digest, 2019b). This means that the existing projects for the most part are still far from practical implementation.

As can be seen from the figure, only the telecommunications and manufacturing industries demonstrate a fairly high level, where almost half of the organizations use software tools for managing production processes. At the same time, the level of implementation of certain technologies that can be attributed to Industry 4.0, for example, RFID technologies, is on average low, as can be seen from Figure 15.3.

In agriculture, despite clear successes and positive examples, there are also difficulties. Among the constraining factors for the implementation of digital solutions, one can single out (Chulok, A., 2021):

- Insufficient infrastructure, which is a traditional problem for regions far from Moscow.
- Regulatory regulation, for example, in terms of obtaining permits for drone flights.
- Lack of specialists with digital competencies.
- The low prevalence of the practice of using digital layouts, mathematical modeling and simulators based on augmented reality technologies for training and advanced training of personnel.

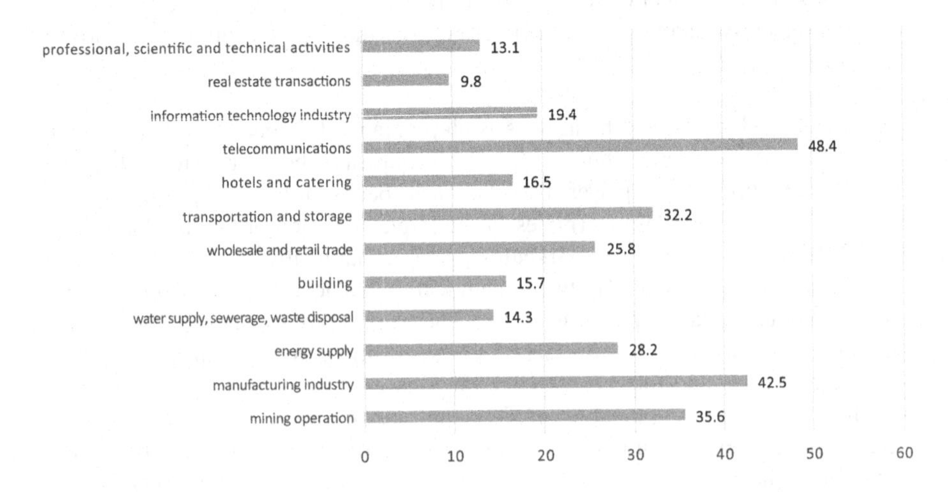

FIGURE 15.2 Use of software for production management by organizations in the entrepreneurial sector, %.

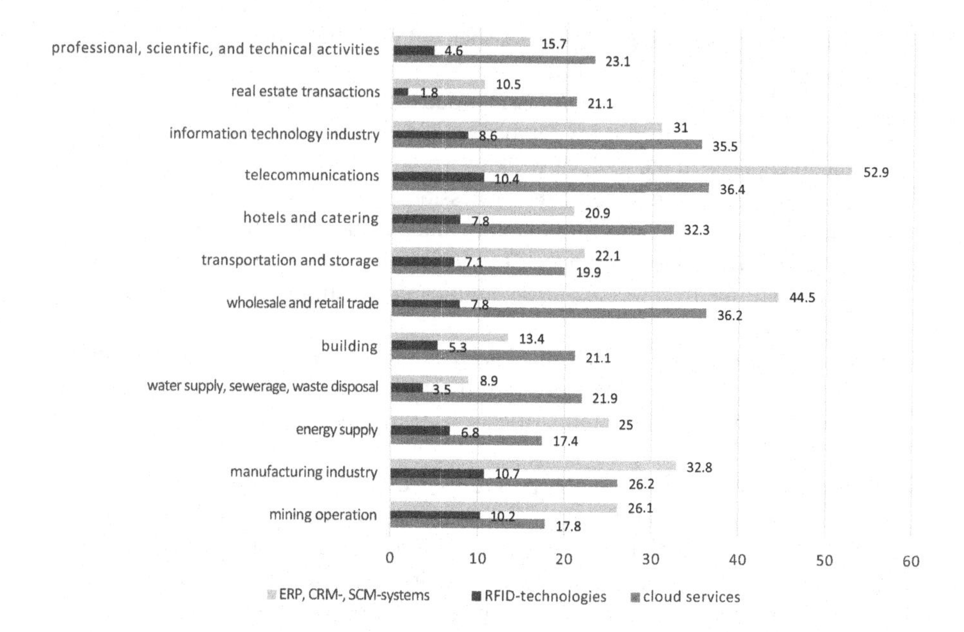

FIGURE 15.3 Digital technologies intensity use by business sector organizations, %.

Studies that surveyed businesses across a wide range of industries show that, in general, companies see the same barriers to digital transformation. More than half of executives cite a lack of quality information and a lack of understanding of the prospects for technology adoption. A quarter notes that there is a lack of understanding of the effect of the introduction of technologies. One in five states that the company lacks financial resources and specialized personnel, as well as a transformation strategy (Sberpro. Media, 2020).

According to the HSE Center for Market Research, 2019 study results, according to 60 percent of enterprise managers, the main factor limiting the digitalization of Russian companies is the lack of financial resources. This problem is typical for SMEs in other countries as well (Horváth & Szabó, 2019; Itsakov et al., 2019). In particular, weak investment activity was noted: almost 60 percent of respondents rated the current level of investment in digital technologies as low. More than 30 percent of enterprises attributed the low readiness of manufacturing industries to digital transformation as a serious limitation. Among the most frequently mentioned barriers (27 percent of respondents) to the digitalization of Russian industry is the lack of favorable and stable economic conditions in the country.

A key factor in the success of digitalization processes is a sufficient number of highly qualified specialists, the availability of appropriate jobs, as well as training systems for specialists with certain competencies for the development and implementation of digital technologies.

The introduction of digital technologies makes significant changes in the requirements for specialists and the needs of companies in personnel:

- Reduced demand for professions associated with the implementation of formalized repetitive operations.
- Shortening the life cycle of professions due to the rapid change in technology.
- Transformation of the set of competencies for some categories of personnel due to changes in the tools of work.
- The emergence of new roles and professions.
- Increasing requirements for flexibility and adaptability of personnel.
- Increased requirements for "soft skills" – the possession of social and emotional intelligence, i.e., those abilities that distinguish a person from a machine.
- Growing demand for specialists with digital dexterity – the ability and desire to use new technologies in order to improve business results (HSE Center for Market Research, 2019; Jacks, Kazantsev, & Serenko, 2020).

In almost 20 percent of cases, the progress of digitalization slowed down due to insufficient experience in technology implementation and a lack of ICT competencies. Slightly less than half (45 percent) of business leaders rated the qualification level of employed specialists in the field of digital technologies as "below normal": the knowledge and skills of the staff were sufficient only to support office software and ICT infrastructure and ensure security and data protection. Freelance specialists were most often involved in making complex decisions related to the support of systems for working with ERP, CRM, HR and databases, and the creation of corporate web portals. Russian universities annually graduate about 25 thousand IT specialists, of which only 15 percent are ready for immediate employment. The average term for a graduate to adapt to the workplace is from 0.5 to 1 year (Association of Electronic Communications, 2017).

To date, the problem of "interpreted artificial intelligence" has not been resolved – automatic systems are not able to give feedback and explain to users the logic of making certain decisions, which is critical in such areas as healthcare, security, law (The Stanford Encyclopedia of Philosophy, 2018), which also makes high demands to specialists performing high-level tasks in the field of business digitalization.

A significant barrier to the development of Industry 4.0 is the problem of ensuring cybersecurity, since traditionally used information security methods are not able to properly protect the digital network. According to research (HSE Center for Market Research, 2019), not all companies apply comprehensive measures to ensure information security.

It is also necessary to highlight the organizational barrier for the development of Industry 4.0 in Russia. Business entities are often not sufficiently aware of the concept and technologies of Industry 4.0, the advantages and difficulties of its implementation; the pace of digitalization at various enterprises is uneven and asynchronous. Some companies are ready for digital transformation in a short time, while others are hampered by the fragmentation of business units and IT departments, outdated infrastructure, which cannot be abandoned or completely replaced in a short time. There is often a mismatch between business goals and the goals of digital business transformation. Top management is often interested in rapid change, but too fast a pace can create unnecessary burden on the company and reduce the efficiency of

TABLE 15.2
Prospects for Digital Transformation in Russia

Opportunities	Drivers	Barriers
1. A qualitative leap in the development of various sectors of the economy. 2. Socio-economic development of the Arctic territories through the creation of data centers.	1. Development of end-to-end digital technologies. 2. State institutions to support digital transformation. 3. Reducing the cost and increasing the availability of digital solutions. 4. Changing market requirements	1. Insufficient experience in the implementation of digital technologies and the lack of a digital transformation strategy for the business. 2. Business lack of financial resources. 3. Lack of highly qualified digital skills.

core business processes. At the same time, too slow a pace of change can reduce the interest in digitalization.

Another significant aspect of the Russian economy is the significant share of the state. Earlier, in the context of digitalization drivers, large-scale and ambitious government programs, which are carried out by giant state corporations, were considered. However, the active influence of the state on the economy leads to a large overregulation of key industries and a slowdown in economic processes, and the situation with 5G is a vivid example of such a slowdown.

Table 15.2 summarizes the opportunities that Industry 4.0 provides to the Russian economy, as well as the drivers and barriers to its development.

15.7 ACKNOWLEDGMENTS

The authors acknowledge the help of Mr. Roman Bogdanov for translation of introduction.

REFERENCES

Abdrakhmanova, G., Vishnevskiy, K., Gokhberg, L. et al., 2019. Что такое цифровая экономика? Тренды, компетенции, измерение. [What Is the Digital Economy? Trends, Competencies, Measurement]. Available from: https://issek.hse.ru/news/261078389.html [Accessed 22 March 2021].

Analytical Center under the Government of the Russian Federation, 2020. Перспективы цифровой трансформации в России. [Prospects for Digital Transformation in Russia]. Available from: https://ac.gov.ru/uploads/5-Presentations/цифровой_трансформации_в_России._Точин.pdf [Accessed 22 March 2021].

Association of Electronic Communications, 2017. Экономика Рунета. Цифровая экономика России 2017. [The Economy of the Runet. Digital Economy of Russia 2017]. Available from: http://raec.ru/upload/files/de-itogi_booklet.pdf [Accessed 22 March 2021].

The Central Intelligence Agency, 2020. Russia. The World Factbook. Available from: www.cia.gov/the-world-factbook/countries/russia/ [Accessed 22 March 2021].

Chulok, A., 2021. Sberpro.Media. Предсказуемое земледелие. Как современные технологии помогают "Белой Даче" повышать производительность. [Predictable Farming. How Modern Technologies Help "Belaya Dacha" to Increase Productivity]. Available from: https://sber.pro/publication/predskazuemoe-zemledelie-kak-sovremennye-tekhnologii-pomogaiut-beloi-dache-povyshat-proizvoditelnost [Accessed 22 March 2021].

Digital Economy Digest, 2019a. Направления развития цифровых технологий в России. [Directions of Digital Technologies Development in Russia]. Available from: https://digest.data-economy.ru/annual-report-2019_Napravleniya_razvitiya_cifrovyh_tekhnologij_v_Rossii [Accessed 22 March 2021].

Digital Economy Digest, 2019b. Актуальность и цели развития цифровых технологий в России. [Relevance and Goals of Digital Technologies Development in Russia]. Available from: https://digest.data-economy.ru/annual-report-2019_Aktualnost_i_celi_razvitiya_cifrovyh_tekhnologij_v_Rossii [Accessed 22 March 2021].

Digital Economy Digest, 2019c. Развитие инфраструктуры интернета вещей. [Development of the Internet of Things Infrastructure]. Available from: https://digest.data-economy.ru/annual-report-2019_Razvitie_infrastruktury_interneta_veshchej [Accessed 22 March 2021].

Federal State Statistics Service, 2020. Промышленное производство в России – 2019 г. [Industrial Production in Russia-2019]. Available from: https://gks.ru/bgd/regl/b19_48/Main.htm [Accessed 22 March 2021].

Gromoff, A., Kazantsev, N., Schumsky, L. and Konovalov, N., 2014. Business Transformation Based on Cloud Services. *2014 IEEE International Conference on Services Computing*, 844–845.

Horváth, D. and Szabó, R. Z., 2019. Driving Forces and Barriers of Industry 4.0: Do Multinational and Small and Medium-Sized Companies Have Equal Opportunities? *Technological Forecasting and Social Change*, *146*, 119–132. Available from: https://doi.org/10.1016/j.techfore.2019.05.021 [Accessed 22 March 2021].

HSE Center for Market Research, 2019. Деловые тенденции и цифровая активность предприятий обрабатывающей промышленности. [Business Trends and Digital Activity of Manufacturing Enterprises]. Available from: https://issek.hse.ru/news/231544472.html [Accessed 22 March 2021].

Institute for Statistical Research and Knowledge Economics, HSE, 2019. Индикаторы цифровой экономики 2019. [Digital Economy Indicators 2019]. Available from: https://issek.hse.ru/ict2019 [Accessed 22 March 2021].

Institute for Statistical Research and Knowledge Economics, HSE, 2020. Индикаторы цифровой экономики 2020. [Digital Economy Indicators 2020]. Available from: www.hse.ru/primarydata/ice2020 [Accessed 22 March 2021].

International Monetary Fund, 2020. *World Economic Outlook Database*. Russia. Available from: www.imf.org/external/pubs/ft/weo/2019/02/weodata/index.aspx [Accessed 22 March 2021].

Itsakov, E., Kazantsev, N., Yangutova, S., Torshin, D. and Alchykava, M., 2019. Digital Economy: Unemployment Risks and New Opportunities. *International Conference on Digital Transformation and Global Society*, 292–299.

Ivanov, A., 2021. Sberpro.Media. Связь, общество и человек: какие инновации ожидают планету. [Communication, Society and Man: What Innovations the Planet Expects]. Available from: https://sber.pro/publication/sviaz-obshchestvo-i-chelovek-kakie-innovatsii-ozhidaiut-planetu [Accessed 22 March 2021].

Jacks, T., Kazantsev, N. and Serenko, A., 2020. Information Technology Issues in Russia. *The World IT Project: Global Issues in Information Technology, World Scientific-Now Publishers Series in Business, Singapore*, *17*, 383–392.

Kinyakina, E., 2020. "Мегафон" начал строить сеть связи в Арктике. Vedomosti. [Megafon Has Started Building a Communication Network in the Arctic]. Available from: www.vedomosti.ru/technology/articles/2020/07/19/834927-megafon-nachal-stroit [Accessed 22 March 2021].

Ministry of Digital Development, Communications and Mass Media of the Russian Federation, 2020a. Паспорт национальной программы "Цифровая экономика." [Passport of the National Program "Digital Economy"]. Available from: https://digital.gov.ru/ru/ministry/common/ [Accessed 22 March 2021].

Ministry of Digital Development, Communications and Mass Media of the Russian Federation, 2020b. Результаты патентного анализа направлений технологического развития цифровой экономики в России и за рубежом. [Results of Patent Analysis of the Directions of Technological Development of the Digital Economy in Russia and Abroad]. Available from: https://digital.gov.ru/ru/documents/7074/ [Accessed 22 March 2021].

Ministry of Digital Development, Communications and Mass Media of the Russian Federation, 2020c. Библиометрический анализ направлений технологического развития цифровой экономики в России и за рубежом с использованием материалов библиометрических баз данных. [Bibliometric Analysis of the Directions of Technological Development of the Digital Economy in Russia and Abroad Using the Materials of Bibliometric Databases]. Available from: https: https://digital.gov.ru/ru/documents/7050/ [Accessed 22 March 2021].

Müller, J. M., Buliga, O. and Voigt, K. I., 2018. Fortune Favors the Prepared: How SMEs Approach Business Model Innovations in Industry 4.0. *Technological Forecasting and Social Change*, *132*, 2–17.

Müller, J. M., Buliga, O. and Voigt, K. I., 2020. The Role of Absorptive Capacity and Innovation Strategy in the Design of Industry 4.0 Business Models-A Comparison Between SMEs and Large Enterprises. *European Management Journal*. Available from: https://doi.org/10.1016/j.emj.2020.01.002 [Accessed 22 March 2021].

National Technology Initiative, 2020. Экосистема национальной технологической инициативы. [National Technology Initiative Ecosystem]. Available from: www.nti.one/nti/eco [Accessed 22 March 2021].

Rostec State Corporation, 2020. Поле возможностей: цифровые решения для сельского хозяйства. [Field of Possibilities: Digital Solutions for Agriculture]. Available from: https://rostec.ru/news/pole-vozmozhnostey-tsifrovye-resheniya-dlya-selskogo-khozyaystva/ [Accessed 22 March 2021].

Sberpro. Media, 2020. Готов ли российский бизнес к цифровой трансформации? [Is Russian Business Ready for Digital Transformation?]. Available from: https://sber.pro/publication/gotov-li-rossiiskii-biznes-k-tsifrovoi-transformatsii [Accessed 22 March 2021].

Skobelev, V., Galimova, N. and Skrynnikova, A., 2020. Совет Федерации предложил создать дата-центры в Арктике. RosBusinessConsulting. [The Federation Council Proposed to Create Data Centers in the Arctic]. Available from: www.rbc.ru/technology_and_media/10/11/2020/5fa93e719a7947e273f617e1 [Accessed 22 March 2021].

The Stanford Encyclopedia of Philosophy, 2018. Artificial Intelligence. Available from: https://plato.stanford.edu/archives/fall2018/entries/artificial-intelligence/ [Accessed 22 March 2021].

Tadviser, 2020. Москва – Умный город. Информационные технологии в Москве. [Moscow – Smart City. Information Technologies in Moscow]. Available from: www.tadviser.ru/index.php/Статья:Москва_Умный_город_(Smart_city)_Информационные_технологии_в_Москве#.D0.9C.D0.BE.D1.81.D0.BA.D0.B2.D0.B0 [Accessed 22 March 2021].

Vilde, O., 2021. Sberpro. Media. Оцифрованный лес: как технологии могут вдвое повысить эффективность лесной отрасли. [Digitized Forest: How Technology Can Double the Efficiency of the Forest Industry]. Available from: https://sber.pro/publication/otsifrovannyi-les-kak-tekhnologii-mogut-vdvoe-povysit-effektivnost-lesnoi-otrasli [Accessed 22 March 2021].

16 Status, Opportunities, and Barriers in Implementing Industry 4.0 in the US

Ling Li and Yang Lu

CONTENTS

16.1 INTRODUCTION

Industry 4.0 disrupts socio-technical ecosystems around the world, as well as in the United States, and has a dramatic social, technological, and regulatory impact on manufacturers. Industry 4.0 is one of ten future projects proposed by Germany in its

DOI: 10.1201/9781003165880-16

"High-Tech Strategy 2020." It is a virtual reality fusion system based on traditional manufacturing, the internet, and the Internet of Things (IoT) to create an intelligent production system. In the future, smart factories will be able to operate on their own, and machines and components will communicate with each other. Cross-industry cooperation will be essential to promote the integration and the development of different industrial sectors, including the logistics and maritime industry, the healthcare sector, the information sector, the communications sector, etc. The fourth Industrial Revolution (4IR) has materialized under the auspices of the Industrial Internet of Things (IIoT), artificial intelligence (AI), machine learning (ML), hyper-converged infrastructure, deep learning, virtualization, and more (Li, 2018, 2020; Xu et al., 2018; Li and Zhou, 2020).

US manufacturers have been one of the most ardent adopters of Industry 4.0. According to a recent markets report, the global IoT in the manufacturing sector was worth $10.45 billion and is predicted to reach $45.3 billion by 2022 in the US (Essentra Components, 2019). As the economy bounces back after the COVID-19 pandemic, investment in the foundational operating technology infrastructure will energize a broad ecosystem of players to leverage the opportunities and benefits of Industry 4.0.

Technology has changed the landscape of manufacturing, service, and global supply chain at a swift pace in the past two hundred years via three industrial revolutions. Companies and organizations that are able to stay ahead of the curve often enjoy a significant advantage over their industry counterparts and competitors. The first three industrial revolutions originated from mechanization, electricity, and information technology. The very first industrial revolution, at the end of the 18th century, created the "industrial age" of machine factories. The second industrial revolution, at the beginning of the 20th century, brought humankind into the "electric age" of mass production. In the middle of the 20th century, computer technology, information technology, and widespread digitalization accelerated the speed of the digital revolution, which is recognized as the third industrial revolution. The invention of computers and the use of programmable machines have expanded human physical strength. The third industrial revolution has initiated a movement to achieve automation through electronics and information technology.

The extensive application of the IoT and service networks in the manufacturing industry has triggered the 4IR, which unfolded at the dawn of the 21st century. This new paradigm is characterized by the fusion of cyber-physical systems, the internet, the IoT (Li, 2020; Posada et al., 2015), industrial information integration (Xu, 2016, 2020) as well as the growing utilization of AI, cloud computing, robotics, 3D printing, data science and, advanced wireless technologies (Sahi et al., 2020; Xu et al., 2018). Under the influence of the 4th industrial revolution, companies will build a global network and integrate machines, storage systems, and production facilities into a cyber-physical system that can complete automatic information exchanges and control actions (Xu et al., 2014; Li, 2018; Lu, 2017).

American companies are digitizing essential functions within their internal business processes, as well as with their supply chain partners along the value chain. Besides, they expand their product portfolio with digital functionalities and introduce innovative, data-based services. Companies are taking actions to effectively increase their overall level of digitization. The US government sees manufacturing as a driving force for economic growth and a vehicle to restore American competitiveness in the world. The US government has lowered corporate tax rates and increased research and development tax credits to stimulate R&D investment. These policies have motivated big companies the drive with full acceleration on the path to achieving the goals of Industry 4.0.

16.2 BACKGROUND OF INDUSTRY IN THE US

Currently, the US is transitioning from traditional factories to smart factories that integrate the IoT and high technology. Industry 4.0 will help the US manufacturing sector and other sectors achieve a higher level of productivity, automation, intelligence, and customization. An analysis of patterns and trends of the performance of American advanced manufacturing conducted by the National Science and Technology Council (NSTC, 2012) revealed a gap that exists between research and development (R&D) activities and the deployment of technological innovations in the high-tech production of goods. This gap has decreased the US's position in the global trade of advanced technology products. Based on the data from the US Census Bureau, the United States ran a trade surplus in the category of advanced technology throughout the 1990s; but by 2010, this surplus had become an $81 billion deficit (NSTC, 2012). American manufacturers were the pioneers in the second and the third industrial revolutions; for example, they commercialized assembly line technology and made automation an industrial standard in the first half of the 20th century, developed industrial robots, and were the first in the globe to utilize this technology in a General Motors plant in 1961. However, now the primary producers of industrial robots are Asian and European manufacturers. The same pattern holds in many other areas of high technology. US-based companies no longer produce color televisions, computer monitors, large rotor disks for turbines, rocket engine parts, missile launch systems, and many other electronic gadgets (Pisano and Willy, 2009). Additionally, the Defense Production Act Committee (DPAC) has identified several vital needed products that are not able to be supplied by reliable and secure domestic producers (NSTC, 2012).

The loss of advanced technology production capabilities affects US national economy. Industry 4.0 heralds an era that the US should restore its competitive advantage in advanced manufacturing and high technology. The acceleration of innovation for advanced manufacturing requires bridging many gaps in the present US industrial sectors, innovation system, and research and development (R&D) activities. In order to keep up with the pace of Industry 4.0, the U. S. government has issued a series of memos and policies to revitalize the manufacturing industry and enhance high-tech innovation. The United States Government introduced the "Advanced Manufacturing Partner Program" in 2011. In 2012, the National Science and Technology Council (NSTC), which is within the Executive Branch of the White House to coordinate science and technology policy across the diverse entities, published the "National Strategic Plan for Advanced Manufacturing" (Lu, 2017; Mittal et al., 2019). This strategic plan formulated by NSTC in 2012 seeks to achieve five objectives: (1) calling for increased investment in advanced manufacturing technology by fostering the more effective use of Federal capabilities and facilities, (2) emphasizing on making the education and training system more responsive to the growing demand for workers who have the skills needed for the advanced manufacturing sector, (3) focusing on creating and supporting national and regional public-private, government-industry-academic partnerships to accelerate the deployment of advanced manufacturing technologies, (4) optimizing government's investment in advanced manufacturing, and (5) increasing the total investments in cutting-edge manufacturing research and development (NSTC, 2012). These objectives are interconnected and interrelated. The policy has addressed the full lifecycle of technology to provide a fertile innovation environment for advanced manufacturing, to enable domestic development of transformative manufacturing technologies, to promote coordinated public and private investment in advanced manufacturing technology infrastructure, and to facilitate rapid scale-up and market penetration of advanced manufacturing technologies (NSTC, 2012).

In the US, small and medium manufacturing firms have formed the backbone of America's manufacturing supply chain. Despite the stagnating progress in manufacturing investment and production due to companies' offshored operations, about 230,000 of US small manufacturing firms have played a key role since 2010 in manufacturing resurgence by adding new manufacturing jobs every year (The US Department of Commerce, 2015). The role of small and medium manufacturers is essential to converting new product ideas to products, and to diffuse new technologies and innovative processes in the manufacturing supply chain. Today, rather than maintaining a vertically integrated supply chain within the company by making most parts and components in company-owned factories, most US large manufacturers depend on outside firms to design and assemble parts. However, small and medium firms usually do not have the resource to invest in advanced manufacturing that is the core of Industry 4.0.

In the report on strengthening small manufacturing firms (The US Department of Commerce, 2015), the authors suggested several measures to reinvest in America's small and medium manufacturers and boost their technological capabilities. In the industry 4.0 era, building tighter linkages between small and large firms is a priority for public and private sectors. The government can leverage its unique assets to help small and medium manufacturers access state-of-the-art research and engineering expertise. Large corporations can provide small and medium suppliers a return on investments if they break new ground in advanced technologies and upgrade their capabilities.

In February 2018, Subcommittee on Advanced Manufacturing, a committee under the National Science and Technology Council, solicited information for National Strategic Plan for Advanced Manufacturing.[1] This effort aimed at asking for inputs from all interested parties on developing a National Strategic Plan for Advanced Manufacturing and improving government coordination for federal programs and activities in support of United States manufacturing competitiveness. The focus is on advanced manufacturing research and development, economic growth, national security, and healthcare improvement. This initiative implies that the US

[1] www.federalregister.gov/d/2018-02160

government tries to enhance government coordination and provide long-term guidance for Federal programs and activities in support of United States manufacturing competitiveness.

16.3 SCIENCE AND TECHNOLOGY STRATEGIC PLAN AND POLICIES FOR ADVANCED MANUFACTURING IN THE US

The intensified global competition has created a sense of urgency for the US government to engage in major industrial revitalization with a core of industrial transformation and development. The White House Office of Science and Technology Policy (OSTP) and the National Science and Technology Council (NSTC) of the US government are responsible for the inter-agency coordination and development of technical reports, strategy documents, and policy memos concerning various scientific and technological topics of importance to the Nation.[2]

Under the Obama Administration, the US government announced the Advanced Manufacturing Partnership plan (AMP), aimed at creating domestic employment opportunities and rebuilding manufacturing competitiveness. In June 2011, the President's Council of Advisors on Science and Technology (PCAST) presented the report, "Ensuring American Leadership in Advanced Manufacturing" and called for a partnership among the government, industry, and academia to identify the most pressing challenges and transformative opportunities to improve the technologies, processes, and products across multiple manufacturing industries (Kuo et al., 2019). A national strategic plan for advanced manufacturing (NSTC, 2012) was then formulated to close gaps among the public, private, universities, and other entities, and address the full lifecycle of the technology. A pilot manufacturing innovation institute was established in 2012 with the Department of Defense (DoD) acting as the lead funding agency based on the suggestion of the report, "Capturing a Domestic Competitive Advantage in Advanced Manufacturing," submitted to President Obama in July 2012 (Kuo et al., 2019). At President Obama's request, more institutes were created in 2014 and 2015 using the lead funding agency authorities and appropriations of the DoD and the Department of Energy (DOE). In December 2014, the US Congress passed the Revitalize American Manufacturing and Innovation (RAMI) Act 8 to create the Network for Manufacturing Innovation Program (NNMI). NNMI Program is the program for coordinating public and private investments to improve the competitiveness and productivity of US manufacturing through the creation of a robust network of manufacturing innovation institutes.

The Trump Administration continued American leadership in manufacturing. In 2018, the National Science & Technology Council in the Office of the President issued a report of Strategy for American Leadership in Advanced Manufacturing. This strategic plan for advanced manufacturing is based on a vision for American leadership in advanced manufacturing across industrial sectors. The vision will be realized by developing and transitioning new manufacturing technologies to market; educating, training, and connecting the manufacturing workforce; and expanding the capabilities of the domestic manufacturing supply chain. Strategic objectives

[2] www.whitehouse.gov/ostp/documents-and-reports/

are identified for each goal, along with technical and program priorities with specific actions and outcomes to be accomplished over the next four years (National Science & Technology Council, 2018).

In 2020, the National Science & Technology Council in the Office of the President put together a document on pioneering the future advanced computing ecosystem. This document outlines a Federal strategic plan for a whole-of-nation approach to pioneering the future national advanced computing ecosystem and establishes the operational and coordination structure to support the implementation of its objectives (National Science & Technology Council, 2020).

According to the World Economic Forum, 65 percent of children entering primary school today will ultimately end up working in completely new job types that currently do not exist (Essentra Components, 2019; Li, 2020). In 2020, the Office of Science and Technology Policy provided Congress a report highlighting the importance of STEM education in the era of Industry 4.0. It encourages the stakeholder community with a window into ongoing and planned Federal activities, intending to lead by example towards the North Star vision of the Federal STEM Education Strategic Plan. This progress report includes a summary of FC-STEM progress on the implementation of the STEM strategy, an analysis of actions developed by the agencies of FC-STEM in support of the Strategic Plan's objectives, a discussion of major focus areas across the Federal STEM education community, a description of the ways Federal agencies will work together to address common challenges, and an inventory of Federal STEM education programs.

These policies related to advanced manufacturing have outlined a Federal strategic vision for a whole-of-nation approach to pioneering the future national advanced computing ecosystem and establishes the operational and coordination structure to

TABLE 16.1
US Manufacturing Policy Tools

	Policy Tools	Policy	Q'ty	%
Supply Side	(1) Public Enterprise		0	0
	(2) Scientific & Technical Development	Advanced Materials (1);	1	1%
	(3) Education	Changing Manufacturing Workforce (3); Better Training for Today's Advanced Manufacturing Workers (1); Educating and Training for Tomorrow's Workers (5);	14	14%
	(4) Information Service	Product Technology Platforms (1); Advanced Manufacturing Processes (1); Data and Design Infrastructure (1); Cross-Cutting Agency Investment (1)	4	4%

	Policy Tools	Policy	Q'ty	%
Environmental Side	(5) Financial	Private-Public Co-Investment (1); Early Procurement (1); Advanced Manufacturing for National Security (1)	3	3%
	(6) Taxation	R & E Federal Credit (1)	1	1%
	(7) Legal Regulatory	Coordinating Federal Investment (5); Raising National Investment in Advanced Manufacturing R & D (3)	7	7%
	(8) Political	Principles and Objectives of the National strategy (4); Strengthening Workforce Skills (5); Creating Partnerships (6); Coordinating Federal Investments (11); Raising national Investment in Advanced Manufacturing R & D (8)	26	25%
Demand Side	(9) Procurement	Early Procurement (1)	1	1%
	(10) Public Services	Accelerating Investment by Small and Medium-Sized Enterprises (4); Strengthening Workforce Skills (14); Creating Partnerships (6); Coordinating Federal Investment (11); Raising National Investment in Advance Manufacturing R & D (8)	43	41%
	(11) Commercial	Advanced Materials (1); Production Technology Platforms (1); Advanced Manufacturing Processes (1); Data and Design Infrastructure (1)	4	3%
	(12) Overseas Agent		0	0

Source: Kuo, C. et al. 2019

support the implementation of Industry 4.0. Kuo et al. (2019) categorized US policy tools into three groups: supply side, environment side, and demand side. Table 16.1 gives a brief description.

The essence of Industry 4.0 is digitization. New ways to design and produce products are created in smart factories that have changed the way companies operate and have transformed humans' role in the labor economy. To unlock the true potential of the cyber-physical system, all business managers along the supply chain should have a digital mindset. In the US, several policies related to digitization have been issued, that include 21st Century Integrated Digital Experience Act (IDEA), Digital Government, and Connected Government Act (Table 16.2).

TABLE 16.2
A Brief List of the Digitalization Policies of the US[3]

Policy	Contents and Explanations
21st Century IDEA (Integrated Digital Experience Act)	"The Act requires all executive branch agencies to modernize their websites, digitize services and forms, accelerate the use of e-signatures, improve customer experience, and standardize and transition to centralized shared services. It requires all government-produced digital products, including websites and applications, to be consistent, modern, and mobile-friendly."
Digital Government	"Building a 21st-century platform to better serve the American people. Strategy principles are information-centric, shared platform, customer-centric, and security and privacy."
Connected Government Act	"The Connected Government Act (H.R.2331) was signed into law on January 10, 2018, and requires new and redesigned federal agency public websites to be mobile-friendly."

16.4 TECHNOLOGIES FOR INDUSTRY 4.0

Digitization, cloud computing, big data, the IoT, and AI are significant factors shaping advanced manufacturing strategy. These technology innovations focus on three areas: (1) the core technologies in the ICT field to create connectivity of factories, warehouse, transportation, and other supply chain entities (2) the enabling technologies that complement core technologies, such as AI and user interfaces, and (3) application domains of these technologies, such as manufacturing enterprise, transportation, and healthcare. The following paragraphs briefly depict technology deployed in implementing Industry 4.0 in the US.

Radio-Frequency Identification (*RFID*). RFID is a wireless automatic identification technology. It can read and write data through wireless identification of the target without identifying the contact between the system and the target in different environments. The advantages are that the operation is fast and straightforward, and that high-speed objects with multiple tags can be identified (Horváth and Szabó, 2019; Müller et al., 2020; Moeuf et al., 2020).

Wireless Sensor Network (*WSN*) is a wireless network composed of many fixed or mobile sensors. They cooperate in a self-organizing and multi-hop manner to sense, collect, process, and transmit the information of the sensed object in the geographic area covered by the network. The information is sent to the network owner. Its advantages are low cost, low power consumption, and miniaturization (Horváth and Szabó, 2019; Müller et al., 2020; Moeuf et al., 2020).

[3] Source: https://digital.gov/resources/checklist-of-requirements-for-federal-digital-services/. Accessed on September 20, 2020.

Artificial Intelligence. AI is a branch of computing known as one of the three cutting-edge technologies in the world. Its direction is the law of human intelligence activities, the construction of artificial systems with a certain degree of intelligence, and the study of fundamental theories and technologies on how to use computers to simulate certain human intelligent behaviors (Horváth and Szabó, 2019; Müller et al., 2020; Moeuf et al., 2020; National Science & Technology Council in the Office of the President, 2019). Gartner estimated that AI applications would create $2.9 Trillion of Business Value in 2021 (Stamford, 2019). Paired with cloud inference, on-device intelligence is a major part of achieving the expected benefits of connectivity across industries. These benefits include user privacy, immediacy, enhanced reliability, and efficient use of network bandwidth. Another intelligence format is the power-efficient AI, which is fundamental across industries and products, spanning from smartphones and automotive to the IoT and data centers. For example, General Electric (GE) has capitalized on AI technology to solve real problems. For instance, leveraging AI solutions to schedule predictive maintenance. Data coming off the gas turbines' sensors are measured and analyzed, and anomalies will be detected for arranging preventive maintenance. Using AI, factories were able to get 20 percent better accuracy in scheduling maintenance which translates into a hundred and sixty thousand dollars per event detected or per failure averted. It sums up to six million dollars per plant per year (Sherry, 2020). The opportunities from a financial perspective for leveraging AI to solve real problems are enormous.

Cloud Computing. Cloud computing is a computing method based on the internet that allows resource information to be provided to other devices on demand. It is characterized by executing distributed calculations on a large number of distributed computers. This approach enables the system to switch resources to the required applications and to access computers and storage systems as needed. Cloud computing provides a reliable and secure data storage center. Users no longer need to worry about data loss, virus intrusion, or other troubles. It provides users with alternatives to using the internet (Horváth and Szabó, 2019; Müller et al., 2020; Moeuf, et al., 2020).

3D Printing. In 2013, 3D printing was developed very quickly. It is broadly used in manufacturing, aerospace and defense, healthcare, and education. Associated with big data, cloud computing, the IoT, AI, and other relevant technologies, 3D will contribute to the development of Industry 4.0. Currently, the US is the largest 3D market in the world. On August 31, 2014, NASA engineers completed their test of the 3D printed rocket launcher. This research was to improve the performance of specific components of the rocket engine. Due to the mixing and reaction of liquid oxygen and gaseous hydrogen in the ejector, combustion temperature can reach 6000 degrees Fahrenheit (about 3,315 degrees Celsius) and can generate 20,000 pounds of thrust (about nine tons). The research verified that 3D printing technology is feasible in manufacturing rocket engines. In 2019, the researchers at the University of California-San Diego used rapid 3D printing technology to create a spinal cord stent that mimics the structure of the central nervous system for the first time, successfully helping rats restore their moving function (Koffler et al., 2019). This innovative application of 3D in healthcare has provided hope for the hundreds of thousands of people worldwide who suffer severe spinal cord injuries each year.

Digital Twin. In 2016, the US Department of Defense, for the first time, proposed the use of Digital Twin technology to maintain and protect aerospace vehicles. The most important inspiration of the digital twin is that it realizes the feedback of the real physical system to the digital model of cyberspace, which is a feat of reverse thinking in the industrial field. Via the use of a digital twin, users are now able to mimic everything that happens in the physical world in the digital space. Entire life tracking with cyclic feedback offers a look at a system's complete life cycle. In this way, the coordination of the digital world and the physical world can be ensured throughout the life cycle. Various simulations, analysis, data accumulation, data mining, and even AI applications based on digital models can ensure its applicability in actual physical systems. This is what digital twin means to intelligent manufacturing. A digital twin is a universally applicable theory and technical method that can be applied across a variety of fields. It is currently used in product design, manufacturing, medical analysis, engineering construction, and other operations.

Robotics. There are many opportunities in the IIoT. Robots have been applied to manufacturing, logistics, even medical and surgery. Today's robots are very good at fixed-function activities, such as picking up a widget from one location and move it to another place. However, robots are not suitable for doing complicated tasks. They are not able to deal with dynamic situations because it is simply too difficult to program a robot to deal with these activities. Nevertheless, this weakness promises a big opportunity for researchers to develop new methods in the future to automate many more functions in the industry that are not currently automatable in robotics.

Digitization. Digitization refers to management activities and methods that use computers, communications, networks, and other technologies to quantify management objects and management behaviors through statistical techniques to produce new products and improve quality. Facing the market competition in smart manufacturing, many companies have realized that the quality of their products and services is the focus of competition. In companies that use traditional quality management methods, the collection and management of quality information are not standardized, and the traceability of quality issues is not easy. Quality decision-making, incomplete analysis information, and other issues restrict the further development of enterprises. With the improvement of the informatization of R&D, production, purchasing, and sales processes, companies hope to use the new quality information system to assist in quality management and achieve informalized communication with other departments (Horváth and Szabó, 2019; Müller et al., 2020; Moeuf et al., 2020).

16.5 DRIVERS FOR IMPLEMENTING INDUSTRY 4.0 IN THE US

American manufacturers and the US supply chain sectors embrace both the present and future Industry 4.0 opportunities. They value real return on investment from Industry 4.0 innovations, technologies, tools, and skills. Business executives see the growth of the bottom line by using these new disruptive technologies and tools to improve productivity and mitigate risk with better quality control. Industry analysts have projected a promising trend of nearly 75 percent of data being created in factories, distribution centers, and retail stores, and about 50 percent of these data being processed, stored, and analyzed directly at the edge.

Several drivers have been identified to motivate US manufacturers to be passionate players in the era of Industry 4.0. They are digitalization and virtual integration, asset utilization, AI penetration, big data collection and analytics, knowledge creation and management, talent development, product and process improvement, and integration of supply chain partners. These drivers are used in the planning, design, and operation of manufacturing and supply chain management, and can be applied to gauge the success of Industry 4.0 implementation. Among these drivers, digitalization and AI technology play a more critical role in extending Industry 4.0 beyond a firm's business boundary to involve global business and trading partners.

The US government sees manufacturing as an engine for economic growth and a vehicle to maintain American leadership status in the world. The US government has lowered corporate tax rates and increased research and development (R&D) tax credits. These policies have motivated big companies the drive with full acceleration on the path to achieving the goals of Industry 4.0.

16.6 SUCCESS CASES

A report published by the European Patent Office in 2017 (EPO, 2017) listed the top 25 companies in the world that made substantial inventions and innovations for the development of Industry 4.0. These 25 companies submitted most pattern applications to the European Patent Office since 2011. The EPO classified more than 48,000 patent applications filed by the end of 2016 into three relevant technology categories of the 4th Industry Revolution, (1) the core technologies in the ICT field that make it possible to create connected objects, (2) the enabling technologies that complement core technologies, such as AI and user interfaces, and (3) application domains of these technologies, such as enterprise, transportation, and healthcare. Six American companies are in the top 25 list. They are Qualcomm, Intel, Honeywell, General Electric, Boeing, and Google.

16.6.1 Qualcomm

Qualcomm Technologies is a leader in technology innovation for the development of Industry 4.0. It delivers robust technologies that support existing and new industries with the power of AI and 5G computing and connectivity solutions. Qualcomm Technologies has introduced a new robotics platform specially designed for power-efficient, high-performance computing robots, and drones for the enterprise. Its new Robotics RB5 platform featuring its QRB5165 robotics processor enables AI, ML, heterogeneous computing, enhanced computer vision, and multi-camera concurrency (Dahad, 2020). Qualcomm's low-complexity, high-performance IoT solutions for manufacturing and distribution centers are leading the way to the next generation of Industry 4.0. The solutions support the factory of the future, bridge the gap between legacy systems with digitalization and reconfiguration of equipment. Additionally, Qualcomm has developed a broad range of solutions for industrial handheld and warehousing automation devices.[4]

---

[4] (www.qualcomm.com)

In 2019, Qualcomm Technologies and Bosch announced a research collaboration on 5G New Radio (NR) for Industrial IoT (Qualcomm, 2019). 5G NR is a new OFDM-based air interface designed to meet an extreme variation of requirements, supporting various devices, services, and spectrum use. 5G NR will be an essential element in the smart, connected factories of Industry 4.0. The collaboration will focus on the applicability of Release 15 5G-NR to existing industrial use cases and the evaluation of the new class of URLLC services to control mission-critical machinery over a wireless industrial Ethernet.

16.6.2 Intel

Intel has a rich history of working with industrial and supply manufacturing chain partners to help achieve workload consolidation. Intel architecture has powered the transformation of people, processes, technologies, and organizations to make Industry 4.0 a reality (Bole, 2019). Intel's AI technologies help leading manufacturers realize the vision of Industry 4.0. Intel, collaborating with ABB Electrification, Alibaba Cloud, Amazon Web Services, Capgemini, Dell Technologies, GE Additive, GE Digital, Hewlett Packard Enterprise, Tridium, Microsoft, and Siemens, are driving the industrial sector forward to realize the goal of Industry 4.0. Businesses utilizing technology with ultra-low latency connectivity over Intel technology-powered 5G and edge networks are able to unlock operational efficiencies and safety improvements in tandem with machine-to-machine (M2M) automation, vision, and AI insights (Bole, 2019).

In 2020, Intel launched Tiger Lake chips, the 11th Gen Intel® Core™ family of processors. A new bar of performance has been established to integrate laptops and internet of things devices. Tiger Lake chips, which include new Intel® Iris® Xe graphics, draw strongly positive press reviews and compete favorably with AMD's Ryzen chip. In the area of 5G technology, with the launch of Snow Ridge, known as the Intel Atom P5900 platform, Intel has become the leading provider for 5G base station silicon by a full year (Intel, 2020). Intel aims to implant AI everywhere. The first satellite with AI on board is now circling the Earth, with an Intel® Movidius™ Myriad™ 2 Vision Processing Unit helping monitor polar ice, soil moisture, and more (Intel, 2020). As AI becomes more pervasive and networks transform to deliver 5G technology, Intel's leaders expect the edge to drive tremendous business value while also improving daily lives.

16.6.3 Honeywell

Honeywell focuses on the application domains of the 4IR technologies and intends to transform the company into an industrial software giant. Using the IIoT to collect data that can then be analyzed, Honeywell develops solutions for other companies to improve performance. For example, airlines can cut costs by lowering maintenance costs and increasing fuel efficiency, and commercial building owners can reduce operating costs by generating energy savings.

In 2020, Honeywell and Tech Mahindra teamed up to build 'Factories of the Future' and build an ecosystem that supports collaboration. The plan aims to leverage industry-leading digital technologies. The two companies lined up to deploy digital transformation, 5G, software capabilities, and engineering expertise to enable manufacturers to scale-up faster. The goal of the collaboration of the two companies is to allow manufacturing customers to expedite their growth and realize the value of Industry 4.0 technologies and solutions (Automotive, 2020).

16.6.4 General Electric

GE has gone through a transformation to be a digital industrial company in the era when industrial information integration is at the center of all actions. Today, GE is a company builds around infrastructure OEMs, water, oil and gas, transportation, medical and operates in key industries such as energy, healthcare, and supply chain. In the last decade, it has grown to be a more than $10 billion software company (LaWell, 2015).

As one of the leading players in the IIoT, GE developed Predix, an IIoT software platform that provides secure edge-to-cloud operational technology and information technology that connect production data, processing, analytics, and services to support industrial customers from GE Digital. Operational technology integrates hardware and software to detect or initiate an alteration through the direct monitoring or control of equipment, processes, or events. Operational technology that GE provides ranges from programmable logical controllers, supervisory control and data acquisition, digital combat simulator, computer numerical control, transportation systems for the built environment, energy monitoring, and more.

In the energy sector, GE has worked with the wind power industry and developed Digital Windfarm technology. This technology uses sensors, data networks, and analytics to customize turbines for maximum efficiency. IIoT sensors attached to GE's power turbines or wind turbines gather data that can be used to monitor and gauge the performance of the physical assets to predict servicing needs better. Renewable energy is energy innovation in the 4th industry revolution (Li, 2020). GE Digital is a key part of its foresight vision and is a $1 billion business by revenue now, with most of it falling in the power and renewable energy segments.

The latest 4IR innovations promise shorter time-to-market, greater flexibility, higher efficiency, and greater quality in the manufacturing and process industries. The most active R&D developers are large companies. Half of all 4IR patent applications between 2011 and 2016 were filed by only 25 companies (EPO, 2017).

Industrial leaders in the US are developing core technologies in the ICT field to create connectivity of enterprises, and develop enabling technologies that complement core technologies, such as AI, and enhance application domains of these technologies. Qualcomm supplies technologies that support existing and new industries with the power of AI and 5G connectivity solutions. Intel designs architecture that has accelerates the transformation of people, processes, technologies, and organizations. By using Industry IoT, Honeywell develops solutions for other companies to improve business performance. Today, GE is a company builds around infrastructure OEMs in key industries such as energy, healthcare, and supply chain. These tech giants are taking the lead to increase the overall level of digitization in the US.

16.7 INDUSTRY 4.0 OPPORTUNITIES

America has an incredible opportunity to become the world leader in the implementation of Industry 4.0 because of its technology and manufacturing prowess. This nation has a large volume of inventions, unparalleled knowledge of managing supply chains, a high concentration of skilled talent, and a vision for advanced manufacturing. The opportunity and the full impact of Industry 4.0 remain to be seen. Currently, the US government has rolled out national policies to develop computing ecosystem integration and implementation, and STEM education programs (Advanced Manufacturing National Program Office, Revitalize American Manufacturing and Innovation, 2014).

The recent figures from the National Association of Manufacturers revealed that the total output of Arizona reached $24.43 billion in 2017, an increase of 8 percent compared to the 2016 figures (Essentra, 2019). The potential of Industry 4.0 has been one of the main drivers. The Greater Phoenix Economic Council commented that 86 percent of manufacturers that include Fortune 500 companies such as Boeing, Honeywell, and Intel believe that they can generate additional revenues through Industry 4.0 techniques. Investments from these major firms have boosted Arizona's economy. Intel announced a $7 billion investment in building the most advanced semiconductor factory in Chandler, Phoenix. The factory will produce microprocessors to power data centers and millions of smart and connected devices such as Augmented Intelligence and advanced transport services (Essentra, 2019). Arizona State University (ASU), located in Phoenix, created the Manufacturing Research and Innovation Hub, the largest additive manufacturing research facility in the south-west to provide cutting-edge plastic, polymer, and metal 3D printing equipment.

Many authors have depicted (Li, 2020; Müller and Voigt, 2018; Müller et al., 2018; Xu et al., 2014; Xu, Xu et al., 2018) the blueprints of a cyber-physical system and opportunities to develop new devices, new production systems, new products, and a new generation of talents. For example, Industry 4.0 will need many template libraries, such as parts library, model library, product library, etc., to form a knowledge base for using design and parameterization to achieve intelligent automation and smart factories. The design model will change from what-you-see-is-what-you-obtain (in which people manually operate the computer to realize the design) to what-you-think-is-what-you-obtain (in which people send instructions through brain waves to construct the plan). Industry 4.0 will be involved in a deep human-computer interaction environment (Müller and Voigt, 2018; Müller et al., 2018).

16.8 BARRIERS IN IMPLEMENTING INDUSTRY 4.0 IN THE US

Industry 4.0 and smart factories are a visionary concept, yet they face a challenging time of being accepted in today's manufacturing plants. There are not many plants that are fully equipped with sensors and computers to automate their production completely. By gaining from digital technology, many businesses fail to fulfill their promises because some of those promises have not been realistic enough (Sherry,

2020). There are several barriers that manufacturers need to understand when they implement Industry 4.0.

16.8.1 Artificial Intelligence

AI technology takes a great deal of human intelligence to be effective. AI is not merely to ask an associate to supply data without knowing what it means. AI should be combined with appropriate industrial expertise and a physical model of machines or processes involved. This is quite different from the commercial internet, where consumers usually do not know the model and the margin for error. Generating a demand forecast with a 10 percent error is a good AI solution for a retailer, but having one air crash for every hundred flights that take-off would be disastrous for an airline company (Sherry, 2020).

16.8.2 Big Data

Simply investing in collecting data would not be enough to bring valuable data to a company. Some data is not useful enough to be collected, while other data should be managed but not stored. To obtain a guaranteed return on investment, firms should start with the tangible improvements that are expected from using 3D printing, computing power from within the cloud, automated data analysis, and more. Then one should find a profitable way to confirm and then extract this value on a wide scale (Sherry, 2020).

16.8.3 Security Issues

In order to realize the security of Industry 4.0, two conditions are essential: one is to ensure that smart factories and industrial products do not cause danger to the environment or personnel, and the other is to prevent data loss and abuse. The virtual and the real world are increasingly integrated, and the challenges facing network information security are becoming more and more severe. Therefore, cloud data need to be protected more effectively (Horváth and Szabó, 2019; Müller et al., 2020; Moeuf et al., 2020).

16.8.4 Standard and Regulatory Issues

Due to the lack of essential software and hardware support for a unified standard, industrial software is usually "bound" by platforms with different technical specifications. Since many things inside and outside the factory are connected to services, a unified reference architecture is needed to describe these standards and promote their implementation (Horváth and Szabó, 2019; Müller et al., 2020; Moeuf et al., 2020). It is not easy to generalize the various facilities involved in Industry 4.0. Government and other authorities still need to generalize appropriate standards and regulations, such as digitalization policies in the US Other related procedures, such as security, IoT operation, and cloud usage, are needed. Without a monitored environment, Industry 4.0 cannot be realized (Horváth and Szabó, 2019; Müller et al., 2020; Moeuf et al., 2020).

16.8.5 System Operation Issues

When the production system is connected with other systems, the management of the entire system becomes complicated. Proper planning, description, and modularization should be established to provide a management foundation for this complex system (Horváth and Szabó, 2019; Müller et al., 2020; Moeuf et al., 2020).

16.8.6 Employee Fear

Many employees will be worried about losing jobs, since Industry 4.0 involves big data, high technologies, and the digitalization process. Organizations or companies need to set up learning and training environment for employees so that they can learn and grasp relevant knowledge and skills to meet the requirements of the development of Industry 4.0 (Horváth and Szabó, 2019; Müller et al., 2020; Moeuf et al., 2020).

16.8.7 Financial Issues

Industry 4.0 is a complicated system that involves so many devices and technologies that a large investment is required to secure the development. Although related policies and strategies have already been constructed in the US, a huge gap between the reality and envision of Industry 4.0 still exists. From a government perspective, more investment and endeavor are necessary for Industry 4.0 to succeed in the US (Horváth and Szabó, 2019; Müller et al., 2020; Moeuf et al., 2020).

16.9 CONCLUSION

With the continuous advancement of science and technology, the world will become smart and intelligent. In tandem with the intensification of globalization and industrial revolutions, technology has undergone significant expansions over the past decades. It is exciting to live in a time of unprecedented transformations and unimaginable technological change. The development of Industry 4.0 will offer tremendous advantages in improving work efficiency and in changing human life.

During the transformation of Industry 4.0, American companies of every size are embracing digital technologies like cloud computing, big data and business analytics, and the IoT. While the benefits of Industry 4.0 are clear, the new Biden administration, the unprecedented pandemic, and the complexity of the journey can make many manufacturers hesitant to embark upon digital transformation. However, some companies are well underway with their transformation efforts to realize long-term success. Companies that take a wait-and-see strategy are already falling behind their peers when it comes to profit and revenue. The time to start the Industry 4.0 journey is now, and it will take time to see the real effects.

Through illustrating the efforts of implementing Industry 4.0 and advanced manufacturing in the US, this article has analyzed the status of the US manufacturing industry, discussed opportunities and barriers to restore American competitiveness through advanced manufacturing in the era of Industry 4.0. Advanced manufacturing is at the crossroad of transformation. While digital transformation is taking shape in nearly every factory, paradoxes can be observed around the industry sectors. Manufacturers are still seeking a path that balances improving the existing production system with the opportunities of ecosystem afforded by Industry 4.0 technologies.

REFERENCES

Advanced Manufacturing National Program Office, Revitalize American Manufacturing and Innovation, 2014. www.manufacturingusa.com/resources/revitalize-american-manufacturing-and-innovation-act. Accessed December 16, 2014.

Automotive, 2020. Honeywell and Tech Mahindra Team Up to Build 'Factory of Future'. February 10, 2020. www.industr.com/en/honeywell-and-tech-mahindra-team-up-to-build-factories-of-future-2470371. Accessed December 30, 2020.

Boles, Christine, 2019. Intel Powering Industry 4.0 for Smart Manufacturing and Data-Centric Transformation. March 28, 2019. https://newsroom.intel.com/editorials/intel-powering-industry-4-0-smart-manufacturing-data-centric-transformation/#gs.omyiy8. Accessed December 30, 2020.

Dahad, N., 2020. Qualcomm 5G and A.I. Robotics Platform Delivers for Industry 4.0 and Drones. June 18, 2020. www.embedded.com/qualcomm-5g-and-ai-robotics-platform-delivers-for-industry-4-0-and-drones/. Accessed December 30, 2020.

EPO.com, 2017. New Patent Study Confirms Growth in Fourth Industrial Revolution Technologies. December 11, 2017. www.epo.org/news-events/news/2017/20171211.html. Accessed December 30, 2020.

Essentra Components, 2019. A Guide to Industry 4.0 in the U.S. January 23, 2019. www.essentracomponents.com/en-us/news/guides/a-guide-to-industry-40-in-the-us. Accessed December 30, 2020.

Horváth, D., and Szabó, R. Z., 2019. Driving Forces and Barriers of Industry 4.0: Do Multinational and Small and Medium-Sized Companies Have Equal Opportunities? *Technological Forecasting and Social Change*, 146, 119–132.

Intel, 2020. 2020 Year in Review. 2020 Intel Corporation.

Koffler, J., Zhu, W., Qu, X., Platoshyn, O., Dulin, J. N., Brock, J., Graham, L., Lu, P., Sakamoto, J., Marsala, M., Chen, S., and Tuszynski, M. H., 2019. Biomimetic 3D-Printed Scaffolds for Spinal Cord Injury Repair. *Nature Medicine*, 25(2), February, 263–269.

Kuo, C., Shyu, J. Z., and Ding, K., 2019. Industrial Revitalization Via Industry 4.0 – A Comparative Policy Analysis Among China, Germany, and the USA. *Global Transitions*, 1, 3–14.

LaWell, Matt, 2015. Matt LaWell Building the Industrial Internet With GE. *Industry Week.com*, October 04, 2015.

Li, L., 2018. China's Manufacturing Locus in 2025: With a Comparison of "Made-in-China 2025" and "Industry 4.0." *Technological Forecasting & Social Change*, 135, 66–74.

Li, L., 2020. Education Supply Chain in the Era of Industry 4.0. *Systems Research and Behavioral Science*, 37(4), 579–592.

Li, L., and Zhou, H. 2020. A Survey of Blockchain with Applications in Maritime and Shipping Industry. *Inf Syst E-Bus Manage*. https://doi.org/10.1007/s10257-020-00480-6.

Lu, Y., 2017. Industry 4.0: A Survey on Technologies, Applications and Open Research Issues. *Journal of Industrial Information Integration*, 6, 1–10.

Mittal, S., Khan, M. A., Purohit, J. K., Menon, K., Romero, D., and Wuest, T., 2019. A Smart Manufacturing Adoption Framework for SMEs. *International Journal of Production Research*, 58(5), pp. 1555–1573.

Moeuf, A., Lamouri, S., Pellerin, R., Tamayo-Giraldo, S., Tobon-Valencia, E., and Eburdy, R., 2020. Identification of Critical Success Factors, Risks and Opportunities of Industry 4.0 in SMEs. *International Journal of Production Research*, 58(5), 1384–1400.

Müller, J. M., Buliga, O., and Voigt, K. I., 2018. Fortune Favors the Prepared: How SMEs Approach Business Model Innovations in Industry 4.0. *Technological Forecasting and Social Change*, 132, 2–17.

Müller, J. M., Buliga, O., and Voigt, K. I., 2020. The Role of Absorptive Capacity and Innovation Strategy in the Design of Industry 4.0 Business Models-A Comparison Between SMEs and Large Enterprises. *European Management Journal*, doi:10.1016/j.emj.2020.01.002.

Müller, J. M., and Voigt, K. I., 2018. Sustainable Industrial Value Creation in SMEs: A Comparison Between Industry 4.0 and Made in China 2025. *International Journal of Precision Engineering and Manufacturing-Green Technology*, 5(5), 659–670.

National Science and Technology Council (NSTC) (2012). A National Strategic Plan for Advanced Manufacturing.

National Science & Technology Council in the Office of the President, 2018. Strategy for American Leadership in Advanced Manufacturing. www.whitehouse.gov/wp-content/uploads/2018/10/Advanced-Manufacturing-Strategic-Plan-2018.pdf. Accessed December 30, 2020.

National Science & Technology Council in the Office of the President, 2019. The National Artificial Intelligence Research and Development Strategic Plan: 2019 Update. www.whitehouse.gov/wp-content/uploads/2019/06/National-AI-Research-and-Development-Strategic-Plan-2019-Update-June-2019.pdf. Accessed December 30, 2020.

National Science & Technology Council in the Office of the President, 2020. Pioneering the Future Advanced Computing Ecosystem: A Strategic Plan. www.whitehouse.gov/wp-content/uploads/2020/11/Future-Advanced-Computing-Ecosystem-Strategic-Plan-Nov-2020.pdf. Accessed December 30, 2020.

Office of Science and Technology Policy, 2020. Progress Report on the Implementation of the Federal Stem Education Strategic Plan. December 2020. www.whitehouse.gov/wp-content/uploads/2017/12/Progress-Report-Federal-Implementation-STEM-Education-Strategic-Plan-Dec-2020.pdf. Accessed December 30, 2020.

Pisano, G. P., and Shih, W. C., 2009. Restoring American competitiveness, *Harvard Business Review*, July–August 2009, page 1–14.

Posada, J., Toro, C., Barandiaran, I., Oyarzun, D., Stricker, D., de Amicis, R., and Vallarino, I., 2015. Visual Computing as a Key Enabling Technology for Industry 4.0 and Industrial Internet. *IEEE Computer Graphics and Applications*, 35(2), 26–40.

Qualcomm, 2019. Qualcomm Technologies and Bosch Announce Research Collaboration on 5G N.R. for Industrial IoT. February 25, 2019, www.qualcomm.com/news/releases/2019/02/25/qualcomm-technologies-and-bosch-announce-research-collaboration-5g-nr. Accessed December 30, 2020.

Sahi, G. K., Gupta, M. C., and Cheng, T. C. E., 2020. The Effects of Strategic Orientation on Operational Ambidexterity: A Study of Indian SMEs in the Industry 4.0 Era. *International Journal of Production Economics*, doi:10.1016/j.ijpe.2019.05.014.

Sherry, Deborah, 2020. If We Want to Build the Factory of the Future, We Need to Go Beyond Industry 4.0. www.ge.com/digital/blog/if-we-want-build-factory-future-we-need-go-beyond-industry-40. Accessed December 30, 2020.

Stamford, Conn, 2019. Gartner Says A.I. Augmentation Will Create $2.9 Trillion of Business Value in 2021. August 5, 2019. www.gartner.com/en/newsroom/press-releases/2019-08-05-gartner-says-ai-augmentation-will-create-2point9-trillion-of-business-value-in-2021. Accessed February 1, 2021.

The U.S. Department of Commerce, 2015. Supply Chain Innovation: Strengthening America's Small Manufacturers. February 2015.

U.S. Industry in Industry 4.0. www.google.com/search?q=US+industry+in+industry+4.0&oq=US+industry+in+industry+4.0&aqs=chrome.69i57j33i22i29i30.10288j0j7&sourceid=chrome&ie=UTF-8.

Xu, L., 2016. Inaugural Issue Editorial. *Journal of Industrial Information Integration*, 1, 1–2, https://doi.org/10.1016/j.jii.2016.04.001

Xu, L., 2020. The Contribution of Systems Science to Industry 4.0. *Systems Research and Behavioral Science*, 37(4), 618–631, https://doi.org/10.1002/sres.2705

Xu, L., He, W., & Li, S., 2014. Internet of Things in Industries: A Survey. *IEEE Transactions on Industrial Informatics*, 10(4), 2233–2243.

Xu, L., Xu, E., and Li, L., 2018. Industry 4.0: State of the Art and Future Trends. *International Journal of Production Research*, 56(8), 2941–2962.

[illegible]

[illegible]

[illegible]

[illegible]

17 Industry 4.0 Experience in SMEs

An International Overview of Barriers, Drivers, and Opportunities

Rubina Romanello, Nikolai Kazantsev, Gilson Adamczuk Oliveira, and Julian M. Müller

CONTENTS

17.1 DIGITALIZATION POLICIES AROUND THE WORLD

Over the years, each country has developed its own digitalization policy, which serves as a national plan to facilitate digitalization and the adoption of Industry 4.0 technologies. Our analysis has highlighted some recurrent policy pillars showing that countries are moving in similar directions to facilitate the upskilling of workers, the development of supporting ICT infrastructures, the digitalization of companies and public governance, as illustrated in Table 17.1 (Appendix). We have generalized some common themes across these national strategic plans:

DOI: 10.1201/9781003165880-17

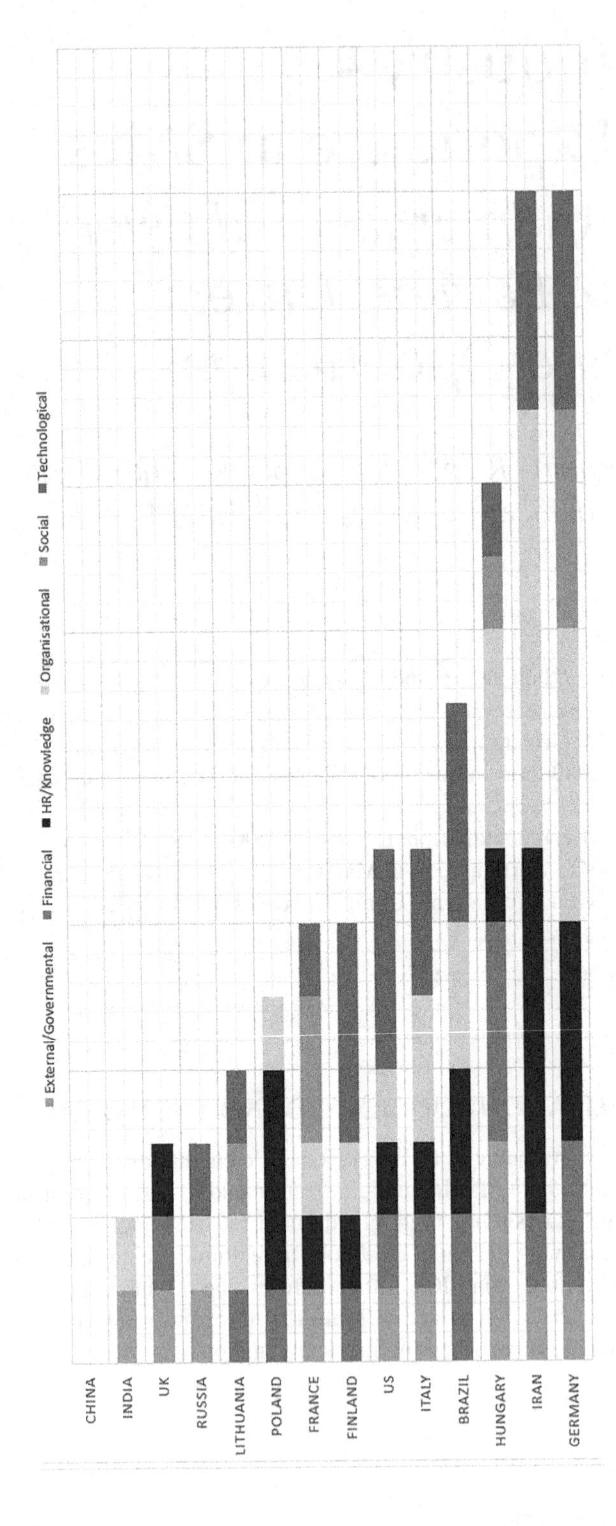

FIGURE 17.1 Graph showing the perceived barriers to implementing Industry 4.0 by SMEs.

1. *Regulatory systems:* Countries invest to improve the regulatory systems in order to solve information privacy issues both for customers and companies, while allowing data sharing and analytics (e.g., GDPR, DMA – European Union countries);
2. *Digital skills:* developing competences related to Industry 4.0 and investments to support training activities (all countries except for Iran). For instance, India has developed the action "Skill India" aimed at upskilling workers to meet labor needs, particularly referring to local SMEs;
3. *Digital infrastructure*: high-speed national broadband (see e.g., Brazil, Finland, Iran, Italy, Lithuania, Russia) and policies aimed at increasing cybersecurity (Brazil, France, Russia);
4. *Co-funding:* access to credits for SMEs and their participation in Industry 4.0 R&D projects to sustain their competitiveness in global value chains (e.g., UK, all European Union countries);
5. *SME orientation*: Brazil, France, Hungary, Italy, Iran, and the US adopted a focus on SMEs, also referring to some specific industries considered strategic for the national economy (e.g., Iran, Poland, and Russia).
6. *Industrial ecosystems:* Most countries recognize manufacturing ecosystems to facilitate knowledge sharing and Industry 4.0 adoption among SMEs (e.g., India – Internet of Things ecosystem, US – advanced computing ecosystem, Hungary, Artificial Intelligence strategy (Finland, Hungary, Lithuania). In addition, countries underlined the importance of facilitating government-academic-industry relations and collaborations (e.g., the US);
7. *Sustainability:* Finland and Germany firstly reconnected tech policies to sustainability goals, leading the way in carbon neutrality strategies.

17.2 INDUSTRY 4.0 BARRIERS

Aligned with extant literature (e.g., Horváth and Szabó, 2019; Müller *et al.*, 2018; Müller and Voigt, 2018), Table 17.2 (Appendix) tries to summarize and classify barriers described in the chapters, based on the categorization of implementation obstacles: Human Resources and Knowledge Factors, Financial Resources, Organizational Factors, Technological Factors, and External and Governmental factors. As illustrated in Table 17.2, each macro-category includes a detailed explanation of micro-categories. Technological barriers mainly related to the limited diffusion of Industry 4.0, which makes difficult to understand which technologies best fit with the SME strategic purpose, to understand how to implement them, to find best practices and to integrate technologies with each other, and with the digital strategy of the company. The lack of digital infrastructure also represents a huge barrier, as particularly SMEs do not have the resources to fill the gap. Organizational barriers mostly related to change management resistance, to the need of reorganizing information and procedures within organization, of outsourcing some Industry 4.0-related activities, to difficulties in finding partners and to the lack of necessary R&D activities. HR barriers mainly related to the lack of digital skills and the need of upskilling workers, whereas the financial barriers relate to the size of investments, the lack of financial support and the unclear returns on the Industry 4.0 investments. Lastly, external barriers refer to the lack of public support, the absence of clear regulations and the competition threat, while social barriers mainly express resistances of union associations or barriers associated with customers' resistances to share data.

- Some countries indicate more barriers rather than others. For this, several possible explanations exist. The result may derive from the cultural reluctance to disclose the country-specific vulnerabilities. However, the features of sampled companies, the perceptions of respondents and the interpretation of researchers also could explain this. However, it is undoubtedly interesting to see that Germany, Iran and Hungary reported the majority of barriers, while China did not report a single barrier.
- Organizational and HR barriers emerged almost at unanimity, as other categories are also recurrent in almost all contexts, except for the social dimension. Future studies could deepen this last aspect, as the lack of social barriers may be interpreted as a latent awareness.
- There is a prevailing trend that for the adopters of Industry 4.0 the several barriers are not technological, for instance, lack of trust, funds, government support. At the same time, the countries who have longer Industry 4.0 experience (such as Western economies), declare more technological barriers (who have experienced Industry 4.0 implementation). In addition, this aspect could be worth a deepening in the future, as it would be interesting to understand whether the Industry 4.0 experience offers a higher degree of consciousness related to the technological difficulties emerging during the adoption and implementation processes.

17.3 INDUSTRY 4.0 DRIVERS

The categorization of drivers follows an extended technology-organization-environment framework, inspired by Horváth and Szabó (2019), as illustrated in Table 17.3 (Appendix). Technological drivers represent a mirror image of technological barriers. In fact, the presence of benchmarks and a well-developed infrastructure, the clear identification of business optimization returns, the possibility to improve or create innovative products and data-decision-making procedures. Organizational drivers relate to a variety of factors and purposes, including business model innovations, decision-making strategy, the entrepreneurial motivation and the strategic goals, digital strategy and culture, the interactions with partners and other actors. HR and knowledge drivers mainly relate to the presence of skilled workforce in the company, whereas the financial drivers refer to the possibility to increase profitability or reduce costs. In the external and governmental category, public support, university-industry collaborations and clear regulatory environments clearly emerge as driving factors. More interestingly, external drivers also relate to competitive pressures arriving from the market/demand and the requirements expressed by customers and suppliers, in line with global value chain dynamics. Requests of technological integration from customers and suppliers clearly represent a driving factor for the implementation of Industry 4.0.

- Brazil, Iran, and Hungary reported the most detailed specification of drivers triggering Industry 4.0 development in their SMEs, while the context of Poland, China and Serbia highlighted relatively less categories of drivers.

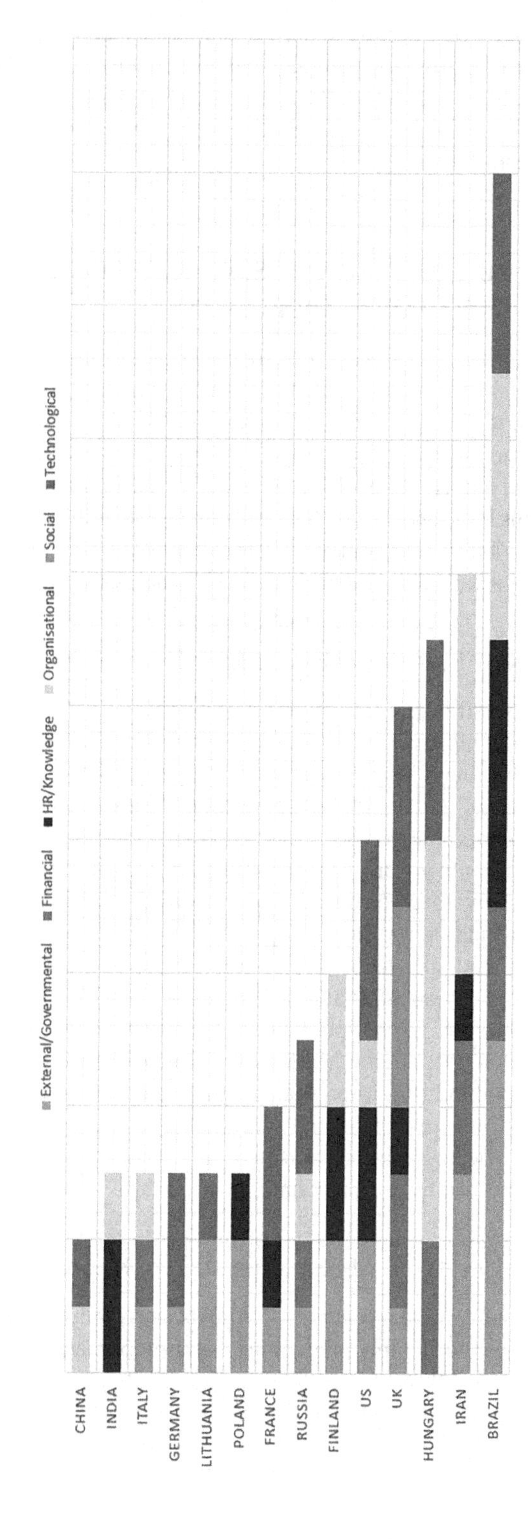

FIGURE 17.2 Graph showing the distribution of drivers to implement Industry 4.0.

Poland particularly underlined external and financial drivers, whereas China and Serbia highlighted technological and organizational drivers.
- Developed countries (e.g., the UK) mention social drivers, which mostly relate to the need of solving societal challenges through the implementation of advanced technologies.

17.4 INDUSTRY 4.0 OPPORTUNITIES, POLICY TARGETS, AND FUTURE RESEARCH DIRECTIONS

We might expect that companies that have already invested in Industry 4.0 keep growing in international markets, whereas a firm that did not invest yet require massive investments to regain competitiveness. Although the costs of technologies would decrease as they spread over the years, still the adoption and implementation process are demanding from the organizational and HR/knowledge perspectives, as highlighted earlier. Based on the arguments of this book and the existing literature, we identified the relevant areas for SMEs related to digital transformation, collaboration in manufacturing ecosystems and the impact on the global value chains (Figures 17.3 and 17.4; Appendix).

17.4.1 Governmental Support to Enable SME Digitalization

Industry 4.0 requires a strategic repositioning of the SMEs, which is a resource-consuming process, bearing errors and financial losses. An inaccurate technological investment for an SME can undermine the solidity and stability of the company, and a failure entails the loss of 'know-how' in production (Müller *et al.*, 2018). As many SMEs operate in networks, the domino effect can cause further capacity shortages in the value chains. National governments should sustain digital transformation and enable SMEs to join new value chains (France, India, Iran, Brazil, and Finland). French and British co-funding programs help SMEs to fill the infrastructural voids. Some countries aid in resolving regulatory framework (India), decrease timing and costs of public services, broadband, and confidentiality infrastructures (European data infrastructure GAIA-X in the EU, Digital infrastructure in Russia), privacy and monopoly issues (GDPR, Digital Markets Act- European Union countries). For instance, in the agri-food industry, precision farming and precision agriculture open the way to improve productivity and the marginality of farmers transformed by Industry 4.0 (Russia). The key issue remains to be able to identify and co-fund sound enterprises, rather than 'zombie firms' (Chang *et al.*, 2021).

17.4.2 SMEs Participation in Manufacturing Ecosystems

The Industry 4.0-driven vertical integration of OEMs usually brings the quick wins. At the same time, SMEs who simply interconnect shop-floor machinery can get their benefit only then Industry 4.0 crosses the firm borders, leading to interconnect different actors in the value chain, including suppliers and customers (Birkel and Müller, 2021; Schmidt *et al.*, 2020; Veile *et al.*, 2020); (Poland, Hungary and USA). The movement to Industry 4.0 concept implies 'lot size of one' and demand-responsive

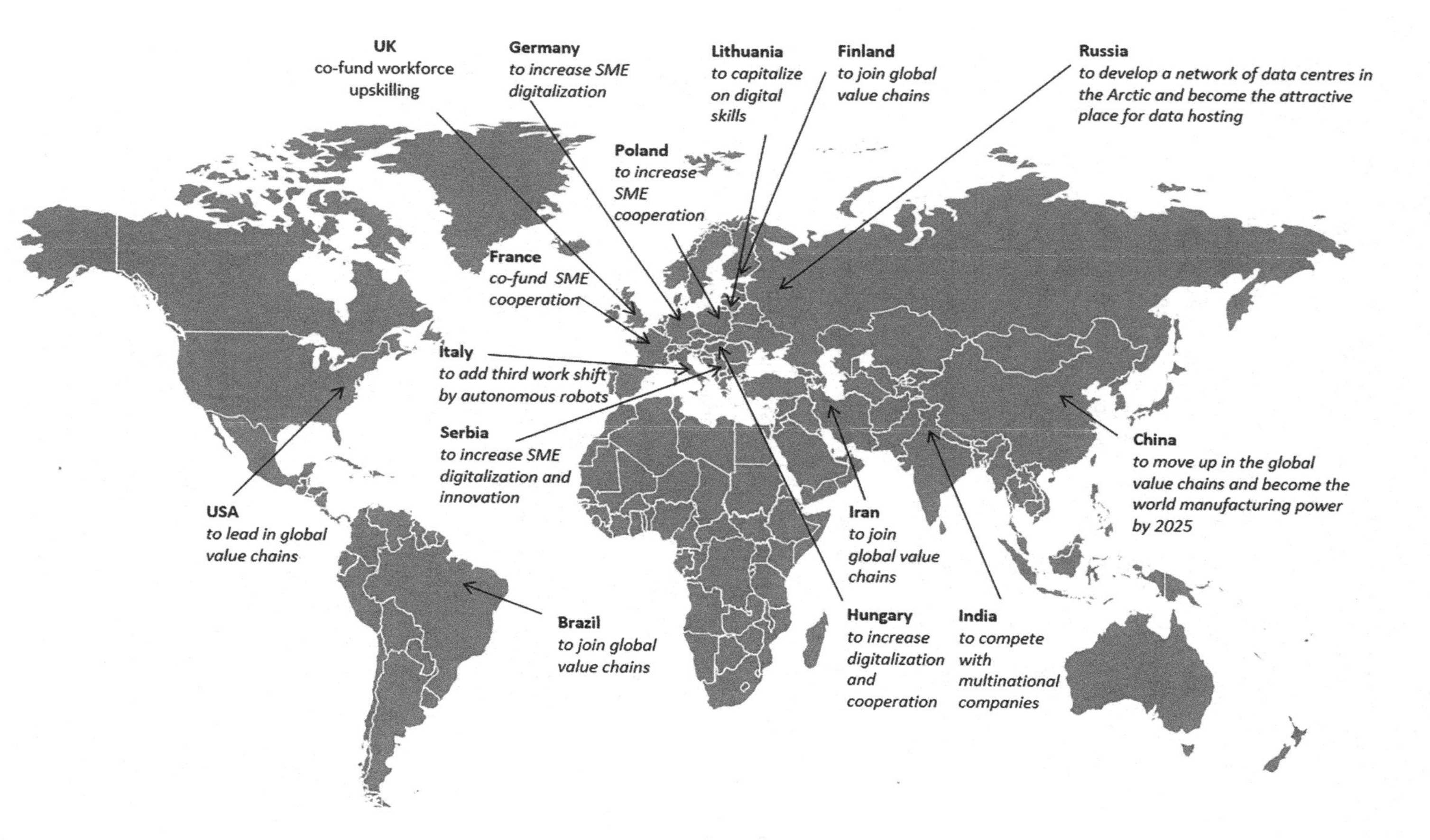

FIGURE 17.3 A top digitalization opportunity for SMEs enabled by the government programs as perceived by the authors based on chapter reviews.

production so that these collaborations formed by SMEs respond to fast changing market needs (Kazantsev *et al.*, 2018). In this vein, SMEs could shorten their path to the customer by joining digital collaboration platforms and collectively fulfill manufacture orders (Cisneros-Cabrera *et al.*, 2021; Kazantsev *et al.*, 2019; Ramzan *et al.*, 2019). Approaching customers via platforms, not via current supply chains, can make SMEs more powerful market participants (France, UK, Germany, and Finland). These dynamics, however, require an openness to collaborations, data collection and sharing, coordinated planning of production and operations, which was shown during the COVID 19-outbreak while developing vaccines. However, most SMEs could be reluctant to share data and be scared of losing control on their core activities (Kazantsev *et al.*, 2018; Müller *et al.*, 2018). Such supply chain partners could limit the operational efficiency and productivity of the whole production network. Although the adoption process in conventional supply chains can be incremental, collaborative R&D activities with universities co-funded by the supra-national bodies, such as European Union (EU), can increase it by improving mutual trust (Bnouhanna and Neugschwandtner, 2019; Müller *et al.*, 2020; Veile *et al.*, 2020).

17.4.3 SMEs International Competitiveness and Impacts

SMEs can strengthen not only their competitive position in the local and regional markets (Müller *et al.*, 2018; Yu and Schweisfurth, 2020), but also in the global value chains (GVCs) (e.g., US, Germany, China). Serving distant markets increases market power of SMEs, allowing them to impact large established organizations (LEOs), who start to lose control over SMEs (Chiarvesio and Romanello, 2018). This can give more strength to SMEs operating in GVCs. For instance, if suppliers integrate with customers, a lock-in effect leads SMEs to convert the current global value chains into a network economy driven by smaller companies. In this vein, SMEs can start collecting data about the products' uses and performances after sales, to enable improvements due to data-driven analytics to improve and innovate products. SMEs that have neither big-data processing facilities nor capacities to store transactional data could supply their data to OEMs or data innovation portals (Rocha *et al.*, 2021), and receive, on exchange, data-driven analytics to optimize their local decision-making (Müller *et al.*, 2020). To store data, SMEs can benefit from partnering with other companies from northern climate zones and lower electricity costs to store transaction data (e.g., Russia) and calling for the world software development regions (e.g., India) for developing apps and data-driven services. This especially is looking at SMEs from developed countries, where the capital and operating costs of supporting data storage infrastructure are high due to the costs of energy resources and additional cooling requirements for server installations. As mentioned in the Russian chapter, there are plans to create powerful data infrastructure in Siberia or the Far East of Russia, but in this case, inter-governmental approvals are required as well as assessments of all possible risks caused by storing data of individuals in the territory of another state (Dobrolyubova *et al.*, 2018). Once legal issues are resolved, SMEs can employ outsourcing and crowdsourcing of analytics based on their data, which will drastically reduce the need in skilled data analyst workforce locally.

17.5 LIMITATIONS

The current misbalance in Internet broadband development limits the optimistic scenario of Industry 4.0 development in SMEs from emerging and Third World countries.[1] As mentioned in the Chinese chapter, 'However, there are big gaps among Chinese enterprises; some of them still need to develop from industry 2.0 to industry 3.0" (see also Müller and Voigt, 2018). Africa and Asia do not widely obtain a stable internet connection; they suffer from higher broadband costs and the need to educate the workforce. As of January 2021, global internet penetration rate by region shows the variation from 26 percent (central Africa) to 96 percent (Northern Europe).[2] The increased cost of IIoT implementation in these regions represent an obvious driver behind the worlds decentralized factory driven by SMEs. It is likely that the developed countries will regulate the market of digitalization services for Third World countries (Dobrolyubova *et al.*, 2018). As the first step, Industry collaboration governance rules are subject to formalization and enforcement (Kazantsev *et al.*, 2018). To this end, there are several initiatives: "Digital Seven" – a network to build digital economy, "promoting open digital standards, knowledge sharing and mutual learning," which includes Canada, Estonia, Israel, New Zealand, South Korea, Uruguay, and the UK.[3] Another example is GAIA-X – a European global project to facilitate the creation of European data and Artificial Intelligence-driven ecosystems to guarantee data sovereignty: elaborate conceptual foundations for shared data infrastructure, create an ecosystem of users and providers, and establish corresponding structures (Bongers, 2020). GAIA-X supports EU-based SMEs in Collaborative Condition Monitoring, Smart Manufacturing, Supply Chain Collaboration in a Connected Industry, Shared Production, and Predictive Maintenance.[4]

17.6 SUMMARY

All the considerations outlined here support the idea that SMEs can benefit from Industry 4.0, which can increase their productivity on the one hand, and enhance the product capabilities and attractiveness to the market on the other hand. Currently, SMEs have not massively adopted the Internet of Things, as the key enabling technology, which was in the past industrial revolutions represented by the steam power, electricity, and computer-aided manufacturing in the world leading economics of the past: Great Britain, the United States, and the Soviet Union. However, the selection and implementation processes are not free of obstacles and policy makers should be aware of their important role in reducing barriers and enhancing drivers of Industry 4.0, also reconnecting with the unique opportunities offered from these technological advancements to SMEs located around the world. Interestingly, SMEs both in advanced economies and in developing countries engaged with the digitization paradigm. However, SMEs from this book still constitute a restricted number compared to the vast majority of companies. As the digital transformation is an inevitable

---

[1] https://ourworldindata.org/grapher/share-of-individuals-using-the-internet (Accessed 26.05.2021)

[2] www.statista.com/statistics/269329/penetration-rate-of-the-internet-by-region/ (Accessed 26.05.2021)

[3] www.digital.govt.nz/dmsdocument/28-d7-charter/html (Accessed 19.05.2021)

[4] www.data-infrastructure.eu/GAIAX/Navigation/EN/Home/home.html

process, for SMEs around the world Industry 4.0 represents a way to remain competitive, also rising barriers to companies that have not digitalized yet, which work with a short-term strategy, become fragile when looking at the medium and long term. Particularly in the context of emerging economies, SMEs that are already in the process of digitization joined with government policies can serve as a catalyst for digitalization of companies in less favored economies. We expect that technological advancement will always become more pervasive within company and across its borders, among the actors of the supply and value chains, requiring a change of mindset in terms of openness to collaborations and co-joint initiatives. Today more than ever, Industry 4.0 forces companies and countries to change their mindset, be more open, co-operate rather than compete and react on the circular economy challenges. To say it in brave words, "*the enemy of the past could be the ally of tomorrow.*"

17.7 ACKNOWLEDGMENTS

The authors wish to thank Mr. Alexey Shvachko for data visualization work (Fig. 1 and 2).

REFERENCES

Birkel, H. S., & Müller, J. M. (2021). Potentials of industry 4.0 for supply chain management within the triple bottom line of sustainability – a systematic literature review. *Journal of Cleaner Production*, 125612.

Bnouhanna, N., & Neugschwandtner, G. (2019). Cross-factory information exchange for cloud-based monitoring of collaborative manufacturing networks. In *2019 24th IEEE International Conference on Emerging Technologies and Factory Automation (ETFA), Zaragoza, Spain*. IEEE, 1203–1206.

Bongers, F. M. (2020). *Three essays on digital and non-digital transformations in business-to-business markets* (Doctoral dissertation, Universität Passau).

Chang, Q., Zhou, Y., Liu, G., Wang, D., & Zhang, X. (2021). How does government intervention affect the formation of zombie firms? *Economic Modelling*, *94*, 768–779.

Chiarvesio, M., & Romanello, R. (2018), "Industry 4.0 Technologies and Internationalization: Insights from Italian Companies," van Tulder, R., Verbeke, A. and Piscitello, L. (Ed.) *International Business in the Information and Digital Age* (*Progress in International Business Research, Vol. 13*), Emerald Publishing Limited, Bingley, pp. 357-378. https://doi.org/10.1108/S1745-886220180000013015

Cisneros-Cabrera, S., Pishchulov, G., Sampaio, P., Mehandjiev, N., Liu, Z., & Kununka, S. (2021). An approach and decision support tool for forming Industry 4.0 supply chain collaborations. *Computers in Industry*, *125*, 103391.

Dobrolyubova, E., Alexandrov, O., Kazantsev, N., & Yangutova, S. (2018). Digital economy: Isolation or collaboration? *Preprint*. www. digicatapult. org. uk.

Horváth, D., & Szabó, R. Z. (2019) Driving forces and barriers of Industry 4.0: Do multinational and small and medium-sized companies have equal opportunities? *Technological Forecasting and Social Change*. Elsevier, 146(October 2018), 119–132. DOI: 10.1016/j.techfore.2019.05.021.

Kazantsev, N., Mehandjiev, N., Sampaio, P., & Stalker, I. D. (2019). A method for facilitating the design of industry 4.0 collaborations and its application in the aerospace sector. DOI: 10.17863/CAM.45896. www.repository.cam.ac.uk/handle/1810/298841 (accessed 26.05.2021).

Kazantsev, N., Pishchulov, G., Mehandjiev, N., & Sampaio, P. (2018). Exploring barriers in current inter-enterprise collaborations: A survey and thematic analysis. In *International Symposium on Business Modeling and Software Design*. Springer, Cham, 319–327.

Kazantsev, N, Sampaio, P., Pishchulov, G., Cisneros Cabrera, S., Liu, Z., & Mehandjiev, N. (2018). A governance metamodel for industry 4.0 service collaborations. In *2018 IEEE World Congress on Services (SERVICES), San Francisco, CA, USA*. IEEE, 47–48.

Müller, J. M., Buliga, O., & Voigt, K. I. (2018). Fortune favors the prepared: How SMEs approach business model innovations in Industry 4.0. *Technological Forecasting and Social Change*, *132*, 2–17.

Müller, J. M., Veile, J. W., & Voigt, K. I. (2020). Prerequisites and incentives for digital information sharing in industry 4.0 – An international comparison across data types. *Computers & Industrial Engineering*, *148*, 106733.

Müller, J. M., & Voigt, K. I. (2018). Sustainable industrial value creation in SMEs: A comparison between industry 4.0 and made in China 2025. *International Journal of Precision Engineering and Manufacturing-Green Technology*, *5*(5), 659–670.

Ramzan, A., Cisneros-Cabrera, S., Sampaio, P., Mehandjiev, N., & Kazantsev, N. (2019). Digital services for industry 4.0: Assessing collaborative technology readiness. In *European, Mediterranean, and Middle Eastern Conference on Information Systems*. Springer, Cham, 609–622.Rocha, C., Quandt, C., Deschamps, F., Philbin, S., & Cruzara, G. (2021). Collaborations for digital transformation: Case studies of industry 4.0 in Brazil. *IEEE Transactions on Engineering Management*. https://doi.org/10.1109/TEM.2021.3061396.

Schmidt, M. C., Veile, J. W., Müller, J. M., & Voigt, K. I. (2020). Ecosystems 4.0: Redesigning global value chains. *The International Journal of Logistics Management*, *ahead-of-print*(ahead-of-print). https://doi.org/10.1108/IJLM-03-2020-0145

Trivelli, L., Apicella, A., Chiarello, F., Rana, R., Fantoni, G., & Tarabella, A. (2019). From precision agriculture to industry 4.0. *British Food Journal*, *121*(8), 1730–1743. https://doi.org/10.1108/BFJ-11-2018-0747

Veile, J. W., Schmidt, M. C., Müller, J. M., & Voigt, K. I. (2020). Relationship follows technology! How industry 4.0 reshapes future buyer-supplier relationships. *Journal of Manufacturing Technology Management*, *ahead-of-print*(ahead-of-print). https://doi.org/10.1108/JMTM-09-2019-0318

Yu, F., & Schweisfurth, T. (2020). Industry 4.0 technology implementation in SMEs – A survey in the Danish-German border region. *International Journal of Innovation Studies*, *4*(3), 76–84.

Appendix

TABLE 17.1
Description of Digitalization Policy Pillars across 14 Countries around the World

	Digitalization Policy Pillars* \ Country	GER	FIN	POL	LIT	FRA	HU	IT	UK	BR	IR	IN	RU	US	CHI
Exogenous	*Legislative environment & regulatory systems*	■	■	■	■	■	■	■	■	■			■		■
	Digital Infrastructure	■	■		■	■	■	■	■	■	■		■		■
	Digital Skills	■	■	■	■	■	■	■	■	■	■	■	■	■	■
	Public administration digitalization		■	■	■		■			■	■	■	■	■	■
	SME orientation					■	■	■	■	■	■			■	
	Direct co-funding	■	■	■	■	■	■	■	■	■	■				■
	Sectoral digital strategies	■		■			■				■				■
	AI strategy	■	■		■		■						■		■
	Cybersecurity strategy	■			■	■				■			■		■
	IoT strategy	■	■				■					■		■	■
Endogenous	*Public-private collaborations*		■			■		■	■		■	■		■	
	International cooperation					■			■	■					
	Sustainability orientation	■	■						■						

Source: Devised by the authors based on the information provided in the chapters. The table aims at providing an overview of the different approaches developed by governments; this representation does not claim to be comprehensive and may not be free of omissions

TABLE 17.2
Description of the Categories of Barriers

Barriers	Micro-categories	Description	Countries
Technological barriers	*Integration*	Relates to the difficulties of integrating different technologies.	BR, IRN
	Digital infrastructure	Relates the lack of digital infrastructure	BR, IRN
	Cybersecurity	Relates to the lack of information on security checks and issues, data security, ownership, and trust in digital ecosystems.	BR, FIN, IRN, US
	Lack of guidance and best practices	Concerns the lack of methodological approaches, best practices, toolkits, the lack of information on system operations and useful data collection.	FIN, US
	Low awareness of technologies	Relates to the low/limited level of awareness and knowledge of Industry 4.0 and of the potential advantages of applying technologies in the different value chain activities.	HU, IT, RU, SERB
	Low level of adoption of technologies	Derives from the limited diffusion of technologies among companies located in the country and the difficulties to assess the digitalization maturity of companies.	FRA, IRN, LIT
	Technologies selection process	Related to the difficulty to find technological solutions for unique needs and selecting the best Industry 4.0 technologies for the company.	BR, IT
Organizational barriers	*Change management*	Employee's fears and resistance to changes in relation to the implementation and utilization of technologies.	IRN, POL, US
	Lack of digital skills	Related to the need to outsource some activities	FRA
	Lack of digitalization strategy	Related to the lack of a clear digital strategy in the company.	BR, FIN, IRN, RU
	Lack of market awareness	Lack of market awareness	HU
	Organizational & codification procedures	Relates to corporate structure and culture, the difficulties deriving from the horizontal integration across the company's boundaries, and the necessity to establish data coding procedures and organizational processes restructuring.	BR, IND, IRN, IT

Barriers	Micro-categories	Description	Countries
	Partner search	Concerning the difficulties to find cooperation and technological partners.	HU, IRN, IT
	Innovation and R&D activities	Related to the absence of innovation management capabilities, limited expenditure in R&D and capacities to manage pilot projects and tests.	HU, LIT, SERB
HR & knowledge barriers	*Lack of digital skills*	Human resources unprepared for digitalization, lack of IT staff, lack of managerial and technological competences to support digitalization, and lack of industrial expertise related to artificial intelligence.	BR, IRN, IT, US
	Skilled labor shortage	Shortage of skilled and specialized workforce.	FRA, HU, POL
	Working class renovation and upskilling	The need to upskill some employees and favor the entry-exit mechanisms related to the workforces.	FIN, IRN, LIT, POL
Financial barriers	*Size of investments*	Related to the high implementation and innovation costs, the lack of internal financial resources, monetary and strategic support, particularly in the case of SMEs.	BR, FIN, HU, IRN, IT, US
	Lack of external funding	Limited access to external funding and venture capital.	HU, LIT
	Unclear returns	Related to the unclear returns on investments, also in relation to the investment size.	BRA, POL, SERB
External & Governmental barriers	*Lack of regulations*	Concerning regulatory barriers and the lack or underdevelopment of laws and regulations on data security, ownership and protection, standardization practices about data format, and clarifications on certification tests.	FRA, HU, IND, IRN, IT, RU, US
	Lack of policy support	Lack of favorable taxation policies.	HU
	Competition and size advantage	Related to the fact that the market is dominated by a few large players that benefit from first-mover advantages.	HU, SERB
Social barriers	*Customer requirements*	Related to social disparities and customers resistance to changes	FRA, HU
	Working associations requirements	Concerning the requirements proposed by trade unions and policies to avoid job losses.	FRA, LIT

Source: Devised by the authors based on the information provided in the chapters; this representation does not claim to be comprehensive and may not be free of omissions

TABLE 17.3
Description of the Categories of Drivers

Barriers	Micro-categories	Description	Countries
Technological drivers	*Benchmarks*	The possibility to confront with other countries, companies and benchmarks on best practices and common errors.	BR
	Business optimization	Immediate increases in productivity and operational effectiveness, in production flexibility, reduced time-to-market, improved monitoring and control of production. Mass customization. Improvements in product quality and project manufacturing capabilities.	CH, HU, IRN, RU, US, UK
	Digital infrastructure	The presence of well-developed digital and AI infrastructures, and rapid increases in computational power.	BR, FR, LI, US, UK, FI
	Product Improvements & Innovation	New products development, materials advancements, optimized and customized services, product and processes innovations.	BR, HU, IRN, US, UK, SERB
	R&D & Innovation activities	High levels of Research and Innovation, scientific and engineering excellence	FR
	Data-driven strategy	Data-driven competitive advantages.	HU
Organizational drivers	*Business model innovation*	The possibility to innovate business models, in terms of servitization for example.	FI, HU
	Strategic decision-making	To make faster and accurate decisions on assets utilizations and ways to reduce cost-performance ratio, also based on real-time data.	CH, HU, US
	Digital strategy & culture	The degree of Industry 4.0 readiness of the company, the presence of a digitalization strategy, the firms' relationships with strategic technological partners and the possibility to start collaborations.	BR, IN
	Inter-organizational collaborations	The exchange of information and interactions with partners and society actors, also related to public-private collaborations.	RU, SERB, BR, FI
	Strategic goals	Better responses to market changes and customers' requirements, repositioning of the firm, new distribution channels, automated services, and agile logistics management, strengthening of corporate image and organizational values, and improved, more attractive workplaces.	HU, IRN, ITA, UK
	Inter-organizational learning	Promotion of Industry 4.0 awareness within the company, through participation at trade fairs and seminars related to digitalization. Knowledge creation and management inside the company.	BR, US

Barriers	Micro-categories	Description	Countries
	Workflow efficiency	Improved procedures and processes of employees at work.	IRN
	Digital strategy & culture	The company digital culture and digitalization approach/strategy.	IN
	Entrepreneurial orientation	Entrepreneurial motivation towards digitalization and risk-taking approach.	FR
HR & knowledge drivers	*Industry 4.0 education & training*	New educational models, training programs, talent management.	BR, US
	New skilled workforce	Hiring Industry 4.0 professionals and highly skilled workforce.	BR, FI
	R&D & innovation activities	The culture of innovation and R&D activities in the company.	FI, PO
Financial drivers	*Entry-level solutions*	Entry-level (low cost) solutions to start implementing Industry 4.0.	BR
	Cost-reduction	Cost-reduction purposes.	HU, IRN, ITA RU
	Profitability	Increases in margins and profitability.	HU; IRN
External /Governmental drivers	*Open technical standards*	Establishment and promotion of open technical standards (interoperability).	BR, UK
	Policy and public support	Governmental and regional programs, Tax reductions and R&D tax credits, and specific credit lines for SMEs and technology implementation.	BR, FR, RU, US
	Innovation cooperation and public-private collaborations	Cooperation in innovative activities, interactions with universities and research institutes, collaboration between the private sector and with governments of other countries to address issues related to data transfer and security.	BR, POL
	Regulatory policy	Establishment of appropriate regulatory frameworks.	BR
	Competitive pressure	High-market competition, market needs, industry competition, increasing economic complexity.	BR, FI, IRN, IT, LI, POL, UK
	Market imbalance/ GVC inputs	International supply chain pressures, responses to customer and supplier requirements, integration of supply chain partners, high exposure to regional and global value chains.	IRN, LI, US, UK
Social drivers	*Social challenges*	Urbanization, circularity, and climate change.	UK

Source: Devised by the authors based on the information provided in the chapters; this representation does not claim to be comprehensive and may not be free of omissions

FIGURE 17.4 A word cloud describing opportunities for SMEs in Industry 4.0.

Printed in the United States
by Baker & Taylor Publisher Services